Builder's
Guide to
Floors

Builder's Guide to Floors

Peter Fleming

McGraw-Hill

New York   San Francisco   Washington, D.C.   Auckland   Bogotá
Caracas   Lisbon   London   Madrid   Mexico City   Milan
Montreal   New Delhi   San Juan   Singapore
Sydney   Tokyo   Toronto

Library of Congress Cataloging-in-Publication Data

Fleming, Peter, 1950–
    Builder's guide to floors / Peter Fleming.
      p.   cm. — (Builder's guide series)
    Includes index.
    ISBN 0-07-021672-X (hc). — ISBN 0-07-021898-6 (sc)
    1. Floors—Design and construction. 2. Flooring. 3. Floor
coverings. I. Title. II. Series.
TH2521.F54 1997
690'.16—dc21
                                        97-26105
                                            CIP

McGraw-Hill

A Division of The McGraw·Hill Companies

2 3 4 5 6 7 8 9 0 FGR/FGR 9 0 2 1 0 9 8 7

ISBN 0-07-021672-X (HC)   0-07-021898-6 (PBK)

The sponsoring editor for this book was Zoe G. Foundotos, the editing supervisor was Bernard Onken, and the production supervisor was Claire Stanley. It was set in Garamond by Jennifer L. Dougherty of McGraw-Hill's professional book group composition unit in Hightstown, N.J.

Printed and bound by Quebecor/Fairfield.

McGraw-Hill books are available at special quantity discounts to use as premiums and sales promotions, or for use in corporate training programs. For more information, please write to the Director of Special Sales, McGraw-Hill, 11 West 19th Street, New York, NY 10011. Or contact your local bookstore.

This book is printed on recycled, acid-free paper containing a minimum of 50% recycled, de-inked fiber.

To my precious wife, Linda, whose loving and constant
encouragement helped turn this thought into reality.
And to my dear son Anthony, whose side I had to leave
for the many hours it took to complete this project.

Contents

5 Ceramic tile   *171*

Preface

Builder's Guide to Floors is designed to provide a fundamental understanding of all the basic types of flooring materials. It describes in detail the following products:

- Carpets
- Hardwood
- Resilient floor coverings
- Ceramic tile
- Laminate floors

It explains not only how these various products are installed, but also how they are manufactured and/or processed. Its purpose is to impart sufficient technical data to anyone wishing to embark upon a career in the floor covering field. Although it demonstrates installation techniques, it is not intended as a do-it-yourself manual. Its aim is to present general guidelines regarding installation procedures so the reader can communicate with clients and tradespeople in an intelligent fashion.

Vast in scope, *Builder's Guide to Floors* is well suited for contractors, installers, or students longing to expand their horizons into this stimulating livelihood. It reveals how to measure rooms and make accurate job site inspections. It describes how to easily sketch a floor plan so as to provide a flooring mechanic with precise shop drawings. It clarifies how to efficiently read a set of architectural blueprints and specifications. Furthermore, it clearly shows how to cost-analyze all the gathered information and present it in a sensible floor covering proposal.

Product knowledge, what to look for when inspecting a job, and comprehension of the various installation systems are what the contemporary floor covering contractor needs to know to be successful in the marketplace today. The flooring industry has opportunities for employment in an exciting and rewarding profession, and proper preparation and education will facilitate entrance into this line of work. *Builder's Guide to Floors* provides an excellent overview of a complex trade. By studying the information presented in this book, the reader will have a firm grasp of how to effectively initiate and operate a prosperous floor covering business.

The closing chapter gives in-depth knowledge on how to actually create a flooring company. Starting with just an idea and four walls, you will learn how to transform those basic elements into a functioning, lucrative operation. Formulating marketing plans, finding a location, keeping accurate records, and generating sales are but a few of the thought-provoking concepts you will be made aware of in this section. Once you are strengthened with this insight, no one can stand in the way of your entrepreneurial dream. Incorporating these principles into a viable, working business plan can lead to financial freedom and unbridled success.

Peter Fleming

Acknowledgments

The process of writing a trade book requires the support of a great many people and organizations. Although it was written in solitude (often in the early morning hours), the completion of this work would not have been possible without the help of everyone involved. Therefore, I would like to take this opportunity to thank all those who contributed to the successful actualization of this project.

First and foremost, I would like to express my sincere thanks to my original editor at McGraw-Hill, April Nolan, for believing in this enterprise from the beginning and for helping me bring this concept to fruition. Also, I would like to thank my current editor, Zoe Foundotos, for turning this manuscript into a completed book.

I would like to thank my father Bartholomew and my mother Rose Fleming for helping me begin my career as a floor covering professional. I would also like to express my appreciation to my many relatives in the flooring industry who have always been a constant source of assistance and inspiration: Genevieve and Frank Muscia, Jr. of Best Buy on Carpets in San Gabriel, California, and Tom Sr., Mary, Tom Jr. and Rose Emerling of Best Carpet Warehouse in Anaheim, California.

In addition, I would like to let the following individuals in Santa Barbara, California, know how appreciative I am of their immeasurable contributions. Brant Elliott for his early editorial work and help in establishing a prudent writing schedule for the timely submission of this manuscript. Lisa Huddelson for her typing and supplementary assistance. Greg Weeks of Holehouse Construction for providing the architectural drawings. Architect Peter Walker Hunt for his technical support. Bob Fatch of Whilt, Fatch and Perry Insurance Services for guidance on insurance coverage. Mahendra Ramawtar and Alice Williams for their photographic expertise. Doug Messick for his superb graphical illustrations. Cynthia Anderson for all her much-needed advice. Frances Halpern for her refreshing stimulation, and most importantly Elizabeth Riley without whom I could never have pulled together all the many little details this project required. She helped with the final typing, editing, illustration requests and manuscript checklist; moreover, she was an exceptional sounding board for the finishing stages of this assignment.

The floor covering industry has also been very supportive of this endeavor and has been more than willing to supply technical data as well as photographic illustrations whenever possible. I am deeply indebted to the following individuals, companies, and associations for their assistance:

- *Armstrong World Industries, Inc.*—James E. Humphrey and Melody Risser
- *Tile Council of America*—Bob Daniels and Ken Erickson
- *Aged Woods, Inc.*—Jeff Horn
- *Bruce Hardwood Floors* and *The Bolton Group*—Dana Agnew and Melissa Austin
- *Construction Services Index (CSI)*—Dominique Fernandez
- *Bellbridge Carpets*—Helma Segal
- *Fritztile*—Stephanie Maher
- *Robbins Hardwood Flooring*—Clark Hodgkins
- *Amorim Revestimentos, S.A., Ipocork Company*—J. Laurentino Couto and Joao Magalhaes
- *National Oak Flooring Manufacturers Association*—Mickey Moore and Patricia Davenport
- *Container Bins, Inc.*—Ms. Chris Goodwin
- *DuPont Corporation*—Gary De Bay, Jack H. Rooke, and Barry Stein
- *James Bunting Advertising* (on behalf of the DuPont Corporation)—Joanne Angelo, John Walker, and Michael Stone
- *The Taunton Press*—Joanne E. Renna (on behalf of R. Bruce Hoadley for use of three of his illustrations)
- *The 3M Company*—Jeffrey L. Hagman
- *Resilient Floor Covering Institute (RFCI)*—Walter Anderson
- *Stern and Associates* (on behalf of the RFCI)—Edith Sachs
- *TEC Incorporated*—Sharon L. Zukley
- *Orcon Corporation*—Pamela Rosen
- *APA—The Engineered Wood Association*—Marilyn Lemoine and Carrie Keeler
- *Carpet and Rug Institute*—Kathryn O. Wise and Brenda Murry
- *Digital Image Studio*—John Ramirez
- *Perstop Flooring*—Lars Von Kantzow
- *Golin Harris* (on behalf of Perstop Flooring)—Kristin Boeke and Joe Micucci

- *Carpet Cushion Council*—Bill Ohler
- *Crossley Carpets*—Teresa Woods
- *Weyerhauser*—Steve Seabert
- *National Tile Contractors Association*—Joe Tarver
- *U.S. Forest Service*
- *Roberts Consolidated Industries*—Walt Zimmerman
- *H.G. Roane Company*—Allen Smith
- *Kentucky Wood Floors Inc.*—John P. Stern
- *Floor Seal Technology, Inc.*—George Donnelly
- *Wool Bureau*—Nina Altschiller
- *Durango Trading Company*—Sharleen D. Daugherty
- *Dal-Tile*—Lisa Harlin
- *Mannington*—Cathy Mansour
- *Stanley-Bostich*—Christopher Dutra and Nancy Lamoriello
- *Solin Design*—Loren Solin
- *Royalty Carpet Mills*—Dennis Egan
- *Tramex, Ltd. Dublin, Ireland*—Sean Fallon
- *American Marazzi*—Massimo Ballucchi
- *Mountain Lumber Company*—Jerome Maddock

"And in the house which king Solomon built for the LORD . . . he
covered the floor
of the house with planks of fir . . . and the floor . . . he overlaid
with gold, within and without."

1 Kings 6:2, 15, 30

1

Flooring basics

All rooms, except perhaps the most Spartan quarters or those used for warehouse purposes, will require some type of floor covering. The modern consumer can chose from a wide array of materials for this floor covering. The kinds of flooring materials available—and their varied styles, textures, and colors—make choosing the appropriate product a challenging and exciting experience. As a floor covering professional, you will be responsible for aiding the consumer in that decision-making process. So, in order for you to advise clients competently, you must have a thorough understanding of floor construction.

There are two primary types of subfloors used in building construction: wood and concrete. Each material has its own dynamic characteristics. Moreover, each substance requires its own special handling. The installation method of a particular floor covering over one surface may be inappropriate for another surface. Therefore, care must be taken when you assess the conditions on a job site. Never assume that a certain substrate exists without making a proper investigation. Consequently, knowledge of fundamental construction methods is indispensable when you are examining any building for future floor covering products.

Floor construction

Each subfloor material (wood or concrete) presents various advantages and disadvantages from a floor covering perspective. Since concrete is generally poured onto the surface of the earth, it is more susceptible to moisture migration and hydrostatic pressure. Wood subfloors, however, are raised off the ground and have a minimum of 18 inches (in) of space between the earth and the subfloor, so they will not have the same water-related problems as concrete slabs. But many structures have both types of subfloors.

1

A great number of homes and businesses may have a concrete slab subfloor on the first floor with a plywood subfloor on the second or succeeding floors. You must treat each situation separately when addressing such issues as preparation work or adhesive choice.

Floor framing for wood floors

Several methods are used for framing houses, including post-and-beam framing and balloon framing. But one of the most commonly used techniques for residential framing is called *platform framing* (Fig. 1-1).

A typical platform frame is built on a foundation wall of concrete or concrete blocks. Into the foundation wall an *anchor bolt* is set (Fig. 1-2). This anchor bolt is a strong metal peg that is firmly embedded in the top of the foundation wall. The anchor bolts secure the *sill plate*. The sill plate is a piece of lumber that has circles drilled out of it, so that the holes fit directly over the waiting anchor bolts. The lumber is secured to the top of the foundation by large nuts that are screwed into the anchor bolts. The sill plate becomes the nailing base for the studs and floor joists.

Once the sill plate is in place, *header joists* are nailed to it. The header joists are nailed perpendicular to the sill plate to provide the vertical height needed to nail in the floor joists. The floor joists are horizontal planks that are laid on their edge side and are nailed into header joists and *girder*. The girder is a large beam made of wood or steel that provides support for the floor. The girder itself is supported by vertical beams underneath the structure. The girder runs perpendicular to the joists and is used as a nailing member for the joists.

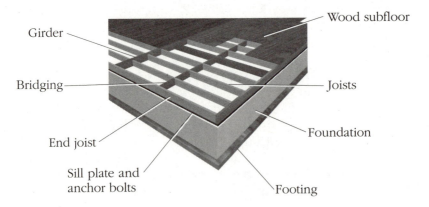

1-1 *Platform framing.* (Courtesy of Doug Messick.)

1-2 *Anchor bolt.*

Bridging is composed of pieces of metal or wood attached between the joists to hold them in position. The bridging helps give the joists their stability and structural integrity. The subfloor is attached to the top of the joists.

The subfloor consists of boards, or panels of plywood or other similar material, that will provide the base for the structure's floor. This subfloor can have carpet attached directly to it, or it can have additional underlayment secured to it if a smoother surface is needed for the floor covering that will follow.

This basic knowledge of platform framing will help in many ways as you begin to examine the different types of floor coverings that can be placed on top of this kind of subfloor. The condition of the subfloor is more critical with hard-surface floorings than with carpeting. Since carpeting is laid over padding (except when it is glued directly to the substrate), carpet tends to hide more imperfections in the subfloor. Hard-surface materials, most notably sheet vinyl flooring, often telegraph any defect in the surface onto which it is applied. In addition, knowledge of subfloor construction is important when you repair damaged floors in existing buildings. Being able to walk on a job, assess the problem, and formulate a solution will be made simpler once you grasp the concept of how subfloors are constructed.

Concrete subfloors

Concrete subfloors for buildings can be poured in a number of ways. The geographic area of the country, and thus the climatic considerations regarding frost and heat, determines the proper technique to be used. In two of the most common methods for constructing concrete subfloors, the foundation and slab subfloor are poured as one unit, or the foundation and slab are poured separately. For the purposes of this discussion, the technique in which the foundation and slab are poured separately will suffice (Fig. 1-3). It is more important for the floor covering specialist to understand the basic properties of concrete than to know the intricate details of foundation and slab construction. This is not to say that you shouldn't familiarize yourself with that information. But pay more attention to concerns for the postconstruction state of the concrete, its alkalinity, and the effects of moisture migration.

The moisture content of a concrete slab is critical in the installation of most floor covering products to its surface. Any time an adhesive is used to secure a flooring material to a concrete subfloor, the ability to bond to that surface is largely influenced by the presence of, or lack of, moisture in the slab. Knowing how water gets into a concrete floor is perhaps the single most important concept a flooring professional should learn about concrete.

Water can enter a concrete slab in a number of ways. It can do so by capillary action, by hydrostatic pressure forces, through the normal curing process, or as upward-moving water vapor from below.

Capillary action (Fig. 1-4) is one of the more common forms of moisture in concrete slabs. Moisture from the water table deep underground will attempt to rise to the surface of the earth, where moisture is released into the atmosphere through evaporation. If, during its ascent, it comes in contact with a concrete slab, it will strive to pass through it. However, if gravel or a polyethylene vapor retarder has been laid underneath the slab, the moisture migration will be inhibited. In its effort to pass through the slab, the water has to saturate the concrete before it can be released. Proper testing (discussed later in this chapter) can show how extensive that saturation may be. Since it is often difficult to determine the exact moisture content of a concrete slab by a visual inspection, it is best to run an established testing procedure prior to installing a floor. An adhesive bond failure could result if the moisture level is too high.

The word *capillary* is used here to describe a slender vein which arises from surface tension in the soil. The moisture travels from the water table through many capillaries up to the concrete slab. In this

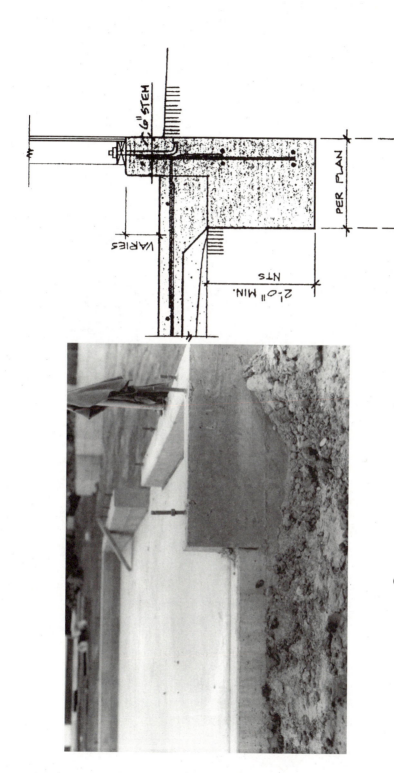

6" STEM

PER PLAN

VARIES

2'-0" MIN.
NTS

1-3 Separate foundation and slab construction.

5

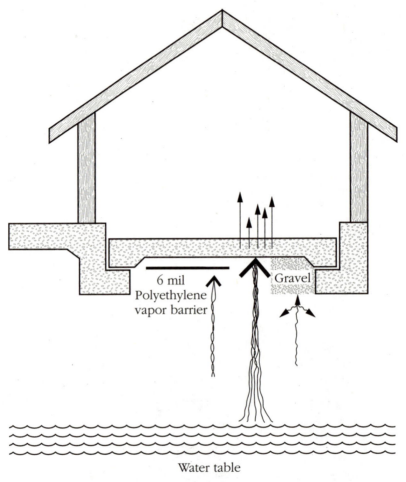

6 mil
Polyethylene
vapor barrier

Gravel

Water table

1-4 *Capillary action.*

action there is a molecular attraction between the liquid and the solid (slab) that causes the water to rise. By their very presence on the soil, slabs tend to attract moisture present in the ground. Voids in slabs, which can be caused by water evaporation during the curing process, can also act as capillaries.

Problems due to hydrostatic pressure are perhaps one of the least likely situations you will encounter as you inspect concrete slabs for moisture content. For hydrostatic pressure to be a contributing factor to any floor covering installation, the concrete slab must be physically below and in direct contact with a momentously high and very saturated source of water. Hydrostatic pressure is measured in pounds per square inch (psi) for liquid water below a certain point in elevation.

For every rise in elevation of 1 columnar foot (ft), the effective pressure is 0.47 psi. (Pressure, psi = $0.47 \times H_c$, where H_c is the column height.) This pressure is the same regardless of volume.

Hydrostatic pressure could occur, for example, if a structure has a basement and an outside water pipe on the side of the building burst and saturated the foundation wall. If the burst pipe went unnoticed for any length of time, the foundation wall would represent a momentously high water source in direct contact with the concrete slab. Excessively heavy rains or poor grading techniques could also trigger a situation in which hydrostatic pressure could develop.

As concrete cures, any excess water is evaporated into the air. A *green slab* (uncured concrete slab) releases the excess water that was used to form the concrete into the movable, workable mixture. The water will continue to leave the slab for a period of several months. As it does so, water leaves voids (capillaries) in the concrete slab. It is these voids that give concrete its porosity. The spaces left after the evaporation process provide the avenues for the future transportation of moisture through the concrete slab. This is precisely why concrete is so susceptible to moisture.

Additionally, as the slab is curing and the moisture is evaporating, it is possible for alkaline salts to be deposited on the surface of the concrete. Alkaline salt deposits, which consist largely of potassium or sodium carbonate, may cause an adhesive bond failure or later may even discolor a sheet vinyl flooring. Test a slab for alkaline salt deposits whenever their presence is likely before you proceed with floor coverings. You can test for alkalinity by using a special pH testing paper or by placing a few drops of 3% phenolphthalein on the surface of the slab. If pH testing paper shows a reading of 10 or higher, or if the drops of solution turn red, the alkalinity must be neutralized. This can be done by mopping the surface with a solution of 4 parts water to 1 part muriatic acid. After this mopping is complete, thoroughly rinse the slab with clean water, allow it to dry, and then retest to ensure that the alkalinity has been neutralized.

Water vapor, which is a combination of air and water, is another source of moisture moving upward through a concrete slab. The variance between the pressures above and below the slab will determine the rate of movement of the vapor through the slab. That movement results when an area of high pressure moves toward an area of lower pressure.

Escaping water vapor becomes a problem for the flooring specialist when the product used on the surface of the slab traps moisture instead of allowing it to escape. Some kinds of flooring products which trap moisture are rubber-backed carpet tiles, rubber tiles, or any material that has a low permeability rate. Whenever a floor covering is installed over a slab that has water vapor passing through it, the potential for moisture-related problems exists. It's possible for a slab to be several years old and appear dry. Yet as soon as a material that traps moisture is installed over the slab, bond failure becomes a distinct possibility.

What can be done to ensure that a concrete slab is sufficiently dry? Perform a moisture test. There are several methods to test for moisture content: mat tests, adhesive bond tests, electronic probe tests, and the calcium chloride tests.

Mat tests are an easy way to check a slab for moisture content. Cut a 24-in × 24-in piece of polyethylene plastic, and tape it directly to the floor with duct tape (Fig. 1-5). Allow the plastic to sit in place for 48 to 72 hours (h). If the slab is releasing moisture, after that amount of time condensation will appear on the underside of the plastic. If moisture is present, the floor requires more time to cure, or else corrective measures must be taken (Fig. 1-6). When you conduct this, or any moisture test, it is always a good idea to check out more than one location in the entire space where you intend to install the new floor coverings. Pay particular attention to areas near walls and posts and in corners.

An adhesive bond test is done by installing several 24-in × 24-in sections of the actual, specified flooring product with adhesive to the surface of the questionable areas. If, after 3 days, the installer must use strength and force to remove the test pieces, the final material will bond to the surface as well. A poor bonding test result means that corrective measures must be taken.

Electronic probes indicate the presence of moisture in the slab but do not measure the water vapor venting through the surface. Consequently, it may be necessary to perform more than one kind of test in a complex situation to get a more precise determination.

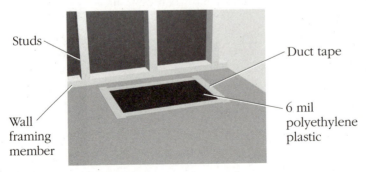

Studs

Duct tape

Wall framing member

6 mil polyethylene plastic

1-5 *Concrete moisture test.* (Courtesy Doug Messick.)

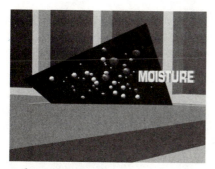

1-6 *Moisture present.* (Courtesy Doug Messick.)

The calcium chloride tests are perhaps the best kind of tests for moisture content in a concrete slab. As a monolithic structure, a slab will have two types of moisture conditions: *static,* which is the actual water present in the slab, and *dynamic,* which is the water vapor emitting from the slab. Calcium chloride tests measure the dynamic condition of a slab floor. By providing quantitative results, this test method should ensure proper adhesion for any *mastics* (adhesives) if the slab passes the test. Please note that it is important to perform this test with close attention paid to the manufacturer's directions on the package.

Choosing the right flooring

Once you have determined that the subfloor is in good condition, you can begin to choose an appropriate floor covering material. Before making a selection, however, you should consider several factors so that the final choice will be the correct one.

Selection guidelines

After you have familiarized yourself with the wide variety of floor covering possibilities, you will find that your choices narrow as you consider your needs and requirements. Keep in mind these guidelines when you are trying to arrive at the proper decision:

- *Comfort.* All floor coverings have a texture and therefore convey a certain degree of comfort. Carpet, being the most luxuriant floor covering, provides the greatest amount of comfort. Hard-surface materials typically serve more utilitarian purposes.
- *Appearance.* Floor coverings, perhaps more than any other single furnishing in a home or an office, set the tone or general overall emotional feeling of a room. Choice of a fabric or material that complements the surroundings greatly enhances the appearance and mood of the space.

- *Wearability.* Since durability is the most important factor in choosing a floor covering, the amount of traffic the floor will receive must be adequately assessed prior to the final selection. For example, kitchens and entryways require a more durable surface than a guest bedroom.
- *Cost.* Every floor covering comes in a variety of grades. In general, the cost is proportional to the quality. Better-quality materials tend to be the most economical in the long run because they provide greater wear and satisfactory service. However, it is important to weigh budgetary considerations when making the final decision. One should strive to buy the best quality possible without overspending.
- *Installation.* Hard-surface floorings have more installation restrictions than carpeting. For this reason, carpeting can cover a multitude of subfloor problems. Consequently, hard-surface materials require more thorough and precise preparation methods. Since the importance of subfloor preparation cannot be overlooked, later sections of this book explain what is necessary to get a subfloor ready to receive new material.
- *Maintenance.* Care and maintenance of a given product will undoubtedly be a prime consideration in choosing a flooring material. The ease of maintaining a modern-day floor covering is of paramount importance to today's consumer. Technological advances in no-wax sheet vinyl flooring and stain-resistant carpeting have paved the way for consumers to spend less time caring for their floors.

Once the above selection guides have been considered, it's time to actually look at the specific types of floor covering choices. Entire chapters of this book are devoted to the various materials available. The most commonly used products examined are carpeting, resilient flooring, hardwood floors, ceramic tile, and new, state-of-the-art floor covering products. Regardless of the product being used, however, each job site must be correctly measured, properly inspected, accurately estimated, and precisely analyzed regarding cost.

Measuring and job inspection tools

You should carry a briefcase that contains all the necessary measuring and job site inspection tools. You must be able to inspect the on-site conditions so as to gain a true understanding of what will be required in both material amounts and preparation work. It is recommended that you use a briefcase because it provides a large, rigid container for your tools. Additionally, it lends an air of professionalism when you walk into someone's house or place of business.

The following are some essential measuring tools (Fig.1-7):

- A 25- or 30-foot (ft) steel retractable tape measure
- A 100-ft steel tape measure for measuring large expanses
- A utility knife for cutting into existing floor coverings
- A linoleum knife for digging and prying open existing flooring
- A pair of pliers
- Screwdrivers
- A hammer
- Wood folding-tape measure
- Pens and pencils
- Pencil sharpener
- Graph paper and clipboard
- Flashlight

Measuring

It is always best to draw a sketch of any rooms to be measured. This is a simple task and takes very little time. A sketch provides you with

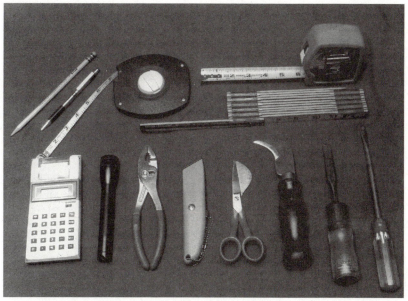

1-7 *Measuring tools.* (Photo by Mahendra Ramawtar.)

a true representation of the job site to which you can refer at a later date. It will also help you to speak intelligently with both the customer and installer regarding specific details of the job.

Drawing a floor plan

Upon entering a job site, visually take in the entire area at a glance. If more than one room is to be measured, view them all before you begin your sketch. This will help you familiarize yourself with the surroundings, and it will assist you in knowing where to begin the drawing on your graph paper. Smaller areas can be started in the middle of the page. Larger areas should be started at the bottom of the page and drawn upward. However, if the rooms lend themselves more to a horizontal interpretation, turn the paper that way and start from the right-hand side of the page and work to the left.

The main objective is to draw as much of the area as possible on one side of one sheet of paper. Otherwise, the spatial relationship of the rooms may not be readily apparent. It is this relationship that is perhaps the most critical aspect of job site measuring, since most products must be installed going in only one direction. It is more difficult to estimate the amount of material needed if you do not know how the rooms are laid out relative to one another. When many rooms are involved, however, use your judgment as to how the rooms should be sectioned off.

Graph paper that has 6 squares per inch is an excellent size for measuring purposes. It is small enough to oblige large drawings, yet not so small as to be illegible. It is not necessary to be so precise that you allow one square for each lineal foot. If you do that, you will spend far too much time counting out each block, and in the process you will lose the essence of what is important. It's not critical to draw everything to scale. What is important is the measurements themselves. If the rooms are drawn slightly askew or out of proportion but the numbers are correct, then the numbers will prevail. It is not suggested that you be sloppy or unconcerned about accuracy; you can, however, loosen up a bit and be more of an artist than an engineer when making a sketch. Scale drawings are fine, but not essential.

For the purposes of this example, the drawing consists of a combination living room and dining room (Fig. 1-8). Examine the drawing carefully, and first try to visualize how the outside walls were drawn. Use your own style to develop your own freehand sketching skills, but here are some suggestions on how to create the drawing.

Begin by drawing the bottom wall (Fig. 1-8*a*) of the living room on a piece of graph paper. Then draw the far right wall (Fig. 1-8*b*). The perpendicular lines at the door opening denote a door that leads

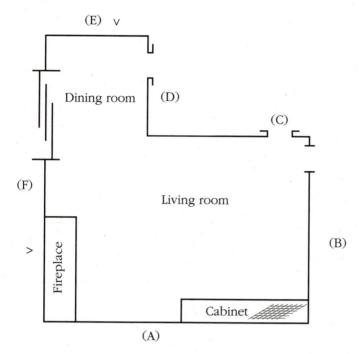

1-8 *Sketch of two rooms, perimeter walls and interior details only.*

to the outside. As you come around parallel (Fig. 1-8c) to the beginning wall, the door opening that has a return mark on it identifies an
inside doorway to another room. It's important to differentiate between the two types of doorways because certain floor molding treatments may be needed to make the transition from one type of
flooring material to another. This information will be needed later
when you are estimating the materials' costs. In this particular instance, the doorway leads into a kitchen, so a metal carpet bar may
be required.

Continue the drawing so that it turns upward toward the second
door that leads into the kitchen (Fig. 1-8d). Next, draw the outside
dining room wall (Fig. 1-8e), and then start back down (Fig. 1-8f) toward the beginning wall. A sliding glass door is indicated on the final
wall by the three parallel lines inside the outside door opening. It is
necessary to make this observation because it will suggest to you certain potential traffic patterns and it will indicate that a great deal of
light will be coming through the glass panes. Finally, finish the drawing by connecting the last wall line back to the original starting point.
As a final touch, sketch in the fireplace and the cabinet. Features such

as these are not chosen just for aesthetic reasons; they clearly show focal areas of the room, and they indicate the source of excess scraps of material (in this case, carpet). Once the drawing is complete, it's time to take the room measurements.

In an area this small, use the 30-ft retractable tape measure. On the graph paper, begin by drawing a mark (>) to show the starting wall of the all measurements that you will take to show the important points in a straight line. Notice that there are two such marks, one for each directional measurement. It is always imperative to get at least two measurements (length and width) for any room or area. The most important information to obtain when you draw a sketch is the point at which there is a directional change within the room. For instance, in Fig. 1-8 the 90° angle created by the directional wall change from the dining room to the living room (Fig. 1-8*d* and *c*) is a critical measurement. Other points to note are the measurements at the doorways. Since you always strive to prevent seams from being perpendicular to a doorway, having the measurements will help when you get to the layout stage of the estimate.

Stretch the tape measure out from wall (Fig. 1-8*f*) to wall (Fig. 1-8*b*). Leave the tape measure lying flat on the floor, and write down all the important directional change points (Fig. 1-9). These measurements should be marked in feet and inches, for example, 1^{10} or 12^3 or 28^3 (for 1 ft 10 in, 12 ft 3 in, or 28 ft 3 in). Mark all those numbers in a straight line with all measurements in sequence until you get to the opposite wall. Moreover, make sure these numbers are all facing the same way. Do not separate little sections and take individual measurements (e.g., do not mark the dining room separately) because you want to see the continuous flow of all the measurements. You can always subtract, add, or calculate certain areas later, once you have the numbers listed sequentially.

Next, measure the area going in the opposite direction from *e* to *a* for the width measurement (Fig. 1-10). Now, when you combine the two drawings, you will see that one set of numbers runs in one direction and the other set runs in the opposite direction (Fig. 1-11). By positioning each set of measurements on its own plane, you can never mistake one directional measurement for another. Whichever > sign the measurements are coming from, they will be on that plane. For example, the fireplace directional change point is 1^{10} from one wall and 16^0 from the other wall. Since the measurements are positioned perpendicular to each other, there can be no confusion about the directions to which they correspond. Once all the measurements are complete, you can use them to lay out a floor plan for the materials to be installed.

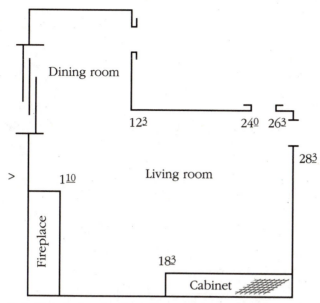

1-9 *Sketch showing length measurements only.*

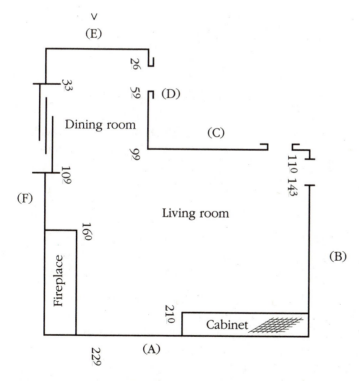

1-10 *Sketch showing width measurements only.*

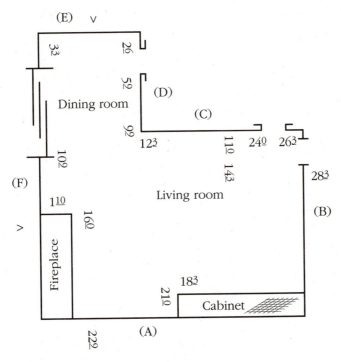

1-11 *Sketch showing both length and width measurements.*

Layout of flooring materials

The layout of floor covering materials involves studying a floor plan drawing and mentally positioning the goods so that the seams will be located in the least obtrusive spots. Since most flooring products that are sold in *roll goods* form cannot be *quarter-turned* (the products cannot be laid perpendicular to each another), usually it takes some planning to arrange the layout effectively. The layout serves several distinct purposes:

- It locates seam placement.
- It helps determine the needed material amounts.
- It aids in the estimation of total job costs.

To properly lay out an area for floor covering products, first it is necessary to know the width of the material and whether there is any pattern match. (Layouts for carpet are no different from those for sheet vinyl flooring or any other roll-type product, so this information can be used for whichever material is chosen.) If the product does have a pattern match, extra material must be added for each additional width (*breadth*) of material that is attached to the first *shot* (length cut) of flooring. Here, let's assume that the product being installed is a solid-colored plush carpet with no pattern match.

Carpet comes in several widths. Typical sizes are 12 ft, 13 ft 6 in, and 15 ft. Of these sizes, the 12-ft width is the most common, so it is used for this example.

First you need to figure out how the 12-ft width of carpet can be best utilized given the floor plan for Fig. 1-11. Look across each set of measurements, and search for key numbers. The number 12, numbers that can be multiplied by to equal 12, and numbers that can divide into 12 equally are key numbers. Therefore, the key numbers in estimating flooring are 3, 4, and 6, because only certain calculations apply: $3 \times 4 = 12$, $2 \times 6 = 12$, $4 \times 3 = 12$, and $6 \times 2 = 12$. Also, $12 \div 3 = 4$, $12 \div 4 = 3$, $12 \div 2 = 6$, and $12 \div 6 = 2$. It may appear as if the above calculations are elementary and unnecessarily trite, but the fact is that when you are trying to lay out floor covering materials using 12-ft-wide products, those calculations are truly the only important ones that can be used effectively. Some examples illustrate this point.

Imagine you have a room that is 18 ft $\times$ 26 ft (Fig. 1-12). Keeping in mind that all materials must be laid with the nap of the carpet running in the same direction, you need the following cuts of carpet:

1. A 12-ft $\times$ 26-ft piece
2. A 12-ft $\times$ 13-ft piece

The logic is you must first cover the entire length of the room with a piece that is 12 ft $\times$ 26 ft. What remains uncovered is a portion that is 6 ft $\times$ 26 ft. Since you cannot quarter-turn the material and run it the short 6-ft distance, and because you can only order the goods in 12-ft widths, you will have to order another piece that is 12 ft wide by some specially determined length. To calculate that length, divide 12 ft (the material width size) by 6 ft (the uncovered width of the room) to arrive at the answer 2. This means that if you took two pieces of 6-ft-wide material, you could fill that uncovered portion. Since you cannot order 6-ft-wide goods, you must order a 12-ft-wide piece and slice it down the middle of the roll, lengthwise, to come up with two 6-ft-wide pieces. You then calculate that 26 ft (room length) divided by 2 (pieces needed to "fill" uncovered portions) equals 13. Therefore, a piece of 12-ft-wide carpet that is 13 ft long, when cut lengthwise down the middle, will produce two 6-ft $\times$ 13-ft pieces that will completely fill the uncovered portion of the room. When these two pieces are joined, they will create a seam known as a *cross-seam*. A cross-seam is perpendicular to the seam that is created by attaching two lengthwise pieces of carpet. You can see in Fig. 1-12 the *fill piece* (material that is used to fill in the remaining portions of a room after at least one complete width has been installed), which is the piece marked 12 ft × 13 ft, and the accompanying other piece marked X^1. The piece designated X^1 is marked thus to show how the fill pieces are to be utilized. It gives an installer a clear picture of how you envisioned laying out the job.

This example can be used for laying out any flooring proposal. Only the multiplication and division choices change. For example, if you had the same room but the room was only 16 ft wide, the calculations would be slightly different (Fig. 1-13). The amount of material needed would be

1. 12 ft × 26 ft
2. 12 ft × 8^2

You would still have 12 ft × 26 ft, but you would only have 4 ft to fill, not 6 ft. So you would calculate that 12 ÷ 4 = 3. It would then follow that the 4-ft area could be filled with three pieces. Then you would divide 26 by 3 to get 8.66. In flooring terms, you would round that number slightly up to 8 ft 9 in to come up with a piece that is 12 × 8^2. One of the main differences here is that there will now be two cross-seams, instead of one cross-seam as shown in the previous example.

Cross-seams are a very interesting subject. It is perhaps the least explained, yet most important topic regarding carpet or any other

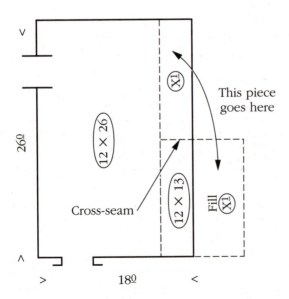

1-12 *Carpet layout for a single room with one cross-seam.*

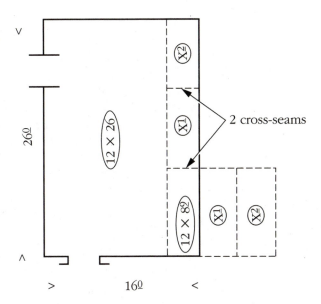

1-13 *Carpet layout for a single room with two cross-seams.*

roll goods product. Cross-seams are created by the necessity to use the least material needed to complete a job. The problem is that these cross-seams can appear very obvious once the job is completed because the material is cut across the rows of the carpet, for example, and reattached later. Carpet is stitched in straight rows, much like a piece of corduroy fabric. If you cut the fabric right down the middle of one of the rows and attempt to reattach it, the fabric should fit back together without any noticeable cut. If, however, you cut that same corduroy fabric across the rows and try to reattach it, the seam will most likely be glaringly obvious. That is the dilemma associated with cross-seams. They are necessary but troublesome. The only way to avoid them is to use full-length cuts whenever possible, but that may often be impractical. In Fig. 1-13, you would have to use two pieces that are 12 ft × 26 ft, but would leave a leftover remnant that is 9 ft × 26 ft, and that is unacceptable because that's an extra 26 square yards (yd^2). But here's the point: The trick in any layout of multiple rooms is to eliminate cross-seams whenever possible by figuring full-length cuts and then utilizing the leftover pieces as fills for other rooms or for closets. This way you use the least yardage, but get the best seam layout. Understanding this last concept is the whole key to effective job estimating.

So, in Fig. 1-11 you have to look at the drawing and the measurements to determine the best starting point from a layout and a traffic standpoint. When you view a floor plan, it is essential to visually establish a starting point and then move 12 ft over (the width of the material) to see where the seams will fall. In the case of Fig. 1-11, the best starting point is at the kitchen door near the rear entry (Fig. 1-14). Since that point is marked at 9 ft 9 in, add 12 ft to that number to arrive at 21 ft 9 in. So the seam drops just beyond the cabinet line, well out of the way of any traffic. The main seam that separates the living room and the dining room is in a high-traffic area, but that is often unavoidable. Moreover, the dining room width is only 9 ft 9 in, so there will be a leftover piece measuring 2 ft 3 in × 12 ft 3 in that can be split in half and used to fill the uncovered portion by the bookshelf, marked X^1 and X^2.

Once the seam locations are determined, you just fill in the length cut amounts. The main cut is 12×28^3, and the dining room cut is 12×12^3. When you add those two cuts, you get 12×40^6 (because you only add the two lengths) which is 54 yd^2 (length times width divided by 9). Now that the yardage is figured out, it's time to make a thorough site inspection before you leave the job.

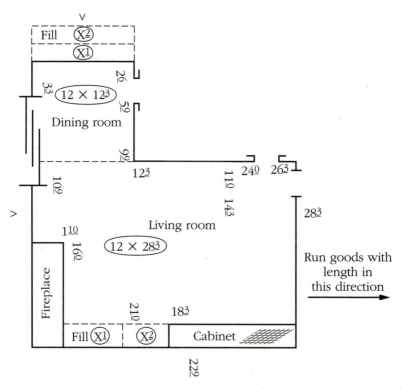

1-14 *Carpet seam layout for two-room sketch.*

Job site inspections

Site conditions reveal key information about current circumstances at a given job location. Research and experience will give you the knowledge necessary to become a good job site inspector. Recognizing what to look for as you walk a job will aid you in identifying probable preparation work, which in turn will allow you to properly analyze all the costs. The ability to locate potential problem areas, and then develop workable solutions for those concerns, will help you determine the special care required to make any subfloor a suitable foundation for the floor covering to follow. Knowing what the current subfloor consists of and what product has been chosen to be installed over it will help make your inspection more fruitful. When you know what product is going to be installed over a certain subfloor, that defines your strategy regarding preparation work. The first thing to determine is the type of the subfloor. Is it a wood subfloor or a concrete slab? Since each subfloor has its own unique characteristics, how you prepare one subfloor is vastly different from how you will prepare another. Also, you do not necessarily inspect a job for wall-to-wall carpet, e.g., in the same way as for resilient sheet vinyl floor covering. So, begin all job site inspections by physically determining precisely the exact type of subfloor. In addition, if there are floor coverings currently on the subfloor, thoroughly investigate those surfaces until you can exactly determine their content. The importance of this physical inspection cannot be overstressed. It may be necessary to actually dig into the existing surfaces, pull back portions of carpeting, or remove transition moldings to peek at what is underneath. Never walk away from a job site inspection without being absolutely certain of current conditions and your approach to any possible problem areas. These are a few of the conditions you will be looking for:

- Subfloor damage or failure
- Existence of any hard-surface floor coverings
- Existence of any old adhesive residue
- Need for removal of any existing flooring products
- Need for any transition moldings, such as reducer strips or thresholds

This is only a short list of the many things you must examine as you walk any job site. Never assume that a certain subfloor exists or, in the case of resilient floor covering, that there is only one layer of flooring. Often, what you may think should be a wood subfloor turns out to be a concrete slab, or what you thought was only one layer of sheet vinyl flooring winds up being two layers. As you become more proficient in your job inspection skills, knowing what to look for will become second nature.

Underlayments and substrates

This is the main question you must ask yourself as you inspect any job: Is the existing subfloor in suitable condition to have a new floor covering installed over it? Once you choose to place a new flooring material over a given surface, you are essentially approving that surface as sufficient and are therefore liable for any resulting problems. Although it may not be your responsibility to make all the necessary repairs, it is your duty to examine the general conditions and bring those conditions to the attention of the owners.

If the subfloor is made of wood, an underlayment may need to be installed prior to the installation of any new floor covering materials. There are a number of underlayment products available on the market, so an understanding of their properties and uses will be most helpful.

Wood underlayments

As previously stated, when you inspect any building for floor covering, it is important to know the kind of material that is to be installed. With just a single layer of plywood installed directly over floor joists, the resulting floor could be suitable for accepting padding and wall-to-wall carpeting; however, it is typically unsuitable to lay resilient floor covering over it. It will lack the structural stability required for glue-down of hard-surface flooring. These subfloors often have nail heads protruding above the surface of the panels. If a hard-surface flooring were installed directly over it, the edges of the plywood and the nail heads might *telegraph* their presence through the material and appear as depressed or raised patterns on the new floor's surface. Therefore, the typical single plywood subfloor is better suited for locations where padding and carpet are to be installed (Fig. 1-15).

Padding and carpet are often an excellent floor covering choice whenever a subfloor is somewhat troublesome. These products have a tendency to hide imperfections in the subfloor, unlike hard-surface floor coverings which require an absolutely smooth underlying surface. (Given that a subfloor for hard-surface materials must be in excellent condition, the following discussion regarding wood underlayments is written as if a full-spread, sheet vinyl resilient flooring is to be installed over it.)

An underlayment over a wood subfloor is generally a wood panel at least $\frac{1}{4}$ in thick. These panels (sheets) typically come in 4-ft × 4-ft or 4-ft × 8-ft sizes. They are made of plywood, or particleboard, oriented

APA RATED STURD-I-FLOOR
(Single Floor)

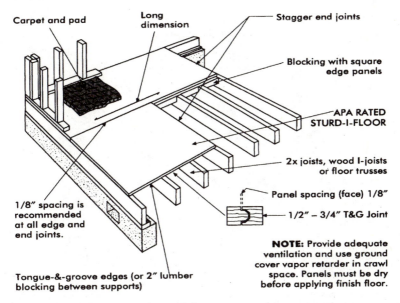

1-15 *Single-layer wood panel floor construction.* (APA–The Engineered Wood Association.)

strand board (OSB), or other artificial composite materials. These panels are installed over the original subfloor (which was attached to the joists), thus making it a double floor (Fig. 1-16). APA (the Engineered Wood Association), a trade organization that represents many of the U.S. wood panel manufacturers, provides recommendations for plywood grades, e.g., that are suitable for use as underlayment (Fig. 1-17). Consult this chart whenever plywood is used as an underlayment.

Underlayment is also used in existing buildings either over old resilient floor covering in lieu of removal or to smooth out a rough surface. Knowing when to recommend a wood underlayment is extremely important. For instance, if there are already two resilient floor coverings on top of each other, most manufacturers of sheet vinyl flooring will not guarantee a third floor on top of the existing two layers. Consequently, you can either remove the first two layers or put down a new $\frac{1}{4}$-in underlayment. Putting down the underlayment does two things. First, you reduce the possibility of harmful substances into the environment from the old flooring material (see Chap. 7); second, you provide the best possible surface for the new flooring. In addition, when you put down new underlayment, the cost can be calculated accurately before the job is started. The cost of

**APA UNDERLAYMENT OVER
APA RATED SHEATHING SUBFLOORING
(Double Floor)**

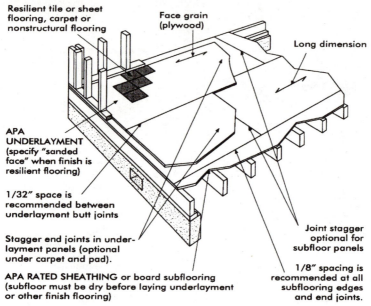

Resilient tile or sheet
flooring, carpet or
nonstructural flooring

Face grain
(plywood)

Long dimension

APA
UNDERLAYMENT
(specify "sanded
face" when finish is
resilient flooring)

1/32" space is
recommended between
underlayment butt joints

Stagger end joints in underlayment panels (optional
under carpet and pad).

APA RATED SHEATHING or board subflooring
(subfloor must be dry before laying underlayment
or other finish flooring)

Joint stagger
optional for
subfloor panels

1/8" spacing is
recommended at all
subflooring edges
and end joints.

1-16 *Double-layer wood panel floor construction.* (APA—The Engineered
Wood Association.)

removal of existing floor coverings, however, is usually billed on a
time-and-materials basis because it is too difficult to accurately pre-
dict how long it will take to pull up the old flooring. Moreover, the
removal process can sometimes gouge or damage the subfloor so
severely that the subfloor is rendered useless. When this happens, an
underlayment is still needed, so the time and money spent on the
removal were unnecessary. When you inspect a job, the goal is to
achieve the best surface for the lowest cost.

APA, the Engineered Wood Association, also rates oriented-strand
board. OSB is panels constructed of strandlike wood particles that are
precisely engineered and bonded together with liquid resins (Fig. 1-18).
These strands are arranged in layers. Each layer is oriented perpendicular to the previous layer, with the surface layers oriented along the
panel's length. By orienting the strands in perpendicular fashion,
strength and structural stability are enhanced.

Particleboard consists of smaller wood particles whose layers are
arranged by the size of the chips. These layers are not oriented with
respect to one another for stability. Particleboard lacks the stability of
many other underlayments, and it is generally not recommended for

TABLE 1

RECOMMENDED PLYWOOD GRADES FOR UNDERLAYMENT

Grade[1][2]	Exposure Durability Classification	Look for These Special Notations in Panel Trademark[3]	Typical Trademarks
APA Underlayment	Exposure 1	Sanded Face	APA THE ENGINEERED WOOD ASSOCIATION / UNDERLAYMENT / 11/32 INCH / GROUP 1 / SANDED FACE / EXTERIOR / 000 / PS 1-95
APA C-C Plugged	Exterior	Sanded Face	APA THE ENGINEERED WOOD ASSOCIATION / C-C PLUGGED / 15/32 INCH / GROUP 1 / SANDED FACE / EXTERIOR / 000 / PS 1-95
APA Underlayment C-C Plugged			APA THE ENGINEERED WOOD ASSOCIATION / UNDERLAYMENT / C-C PLUGGED / GROUP 1 / SANDED FACE / EXTERIOR / 000 / PS 1-95

1-17 *Recommended plywood grades for underlayment.* (APA—The Engineered Wood Association.)

APA A-C	Exterior	Plugged Crossbands Under Face(4)
APA B-C	Exterior	"
APA A-D	Exposure 1	"
APA B-D	Exposure 1	"

APA
THE ENGINEERED
WOOD ASSOCIATION

A-C 15/32 INCH
GROUP 1

PLUGGED CROSSBANDS UNDER FACE

EXTERIOR
000
PS 1-95

APA Underlayment A-C	Exterior	Sanded Face
APA Underlayment B-C		

APA
THE ENGINEERED
WOOD ASSOCIATION

UNDERLAYMENT

B-C 11/32 INCH
GROUP 1

SANDED FACE

EXTERIOR
000
PS 1-95

(1) Veneer-faced, 19/32-inch or thicker panels; or APA Rated Sturd-I-Floor, Exposure 1 or Exterior marked "Sanded Face"; or APA Marine Exterior plywood also may be used for underlayment under resilient floor covering.

(2) Specific plywood grades and thicknesses may be in limited supply in some areas. Check with your supplier before specifying.

(3) Recommended for use under resilient floor covering.

(4) "Plugged Crossbands (or core)," "plugged inner plies" or "meets underlayment requirements" may be indicated as alternate designation in or near trademarks.

1-17 *Continued*

Engineered To Get The Most From Natural Wood.

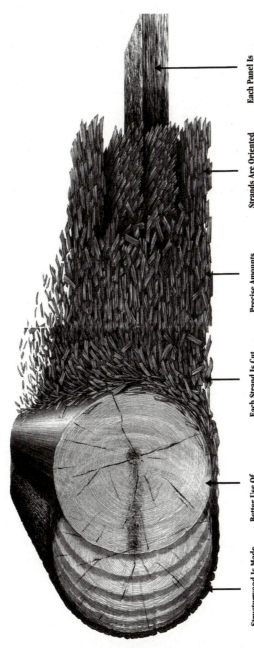

Structurwood Is Made From Quality Logs.
Weyerhauser uses logs from quickly regenerating wood species.

Better Use Of Natural Resources.
The bark is removed so only the best part of the log is used. All of the remaining material is used to make the finished panels. This efficient use of our precious natural resources is in keeping with our philosophy of responsible environmental stewardship.

Each Strand Is Cut To Exact Dimensions.
A series of specially designed knives cut the log into uniform strands. Each strand follows the original grain of the wood, which takes advantage of the inherent strength of the fiber, resulting in a product of superior strength and stiffness.

Precise Amounts Of Wax And Resin Coat Each Strand.
Computer controlled amounts of wax and resin are applied to the strands. This state-of-the-art application ensures uniform bonding and resistance to moisture.

Strands Are Oriented Cross-Dimensionally For Superior Strength.
Each of the four layers of strands is oriented within a tight range of axis to maximize panel strength, stiffness and stability.

Each Panel Is Engineered To Perform To End-Use Specifications.
Exceptional quality control assures that each panel is 100% uniform and consistent. This consistency of quality reduces waste and assures each panel is precisely manufactured for its particular end use.

1-18 *Oriented-strand board.* (Weyerhauser Company.)

full-spread sheet vinyl installations. It is usually only suitable for the perimeter glue installations in which a 4- to 6-in band of adhesive is spread around the perimeter of the room and where needed for seams.

Some of the new composite panels offer excellent performance warranties and are well worth looking into. These products will be subject to the same performance requirements of the flooring manufacturers. So before you select an underlayment, check with the floor covering supplier for the most appropriate product. It is your job to recommend the most suitable underlayment for a given job because once it is laid down, it's your responsibility if the choice is contrary to the flooring manufacturer's suggestions. If there is a failure of the underlayment product, you may be able to make a claim against the manufacturer of the underlayment product if the manufacturer clearly recommended that particular product for your purpose.

Concrete substrates

A concrete floor should be smooth, sound, and free of moisture. In addition, it should have no imperfections and irregularities. Inspecting jobs that have concrete subfloors requires different evaluation procedures from those for wood subfloors. As discussed earlier, moisture content considerations are extremely important. Do appropriate testing and carry out follow-up methods for moisture content. Also, since the surface is rigid and will not accept a wood underlayment, any leveling or smoothing must be done with an appropriate liquid latex underlayment that is spread onto the floor. Follow the manufacturer's directions for using a liquid underlayment product.

Any expansion joints or cracks should be filled with an acceptable patching compound. Point out any areas that may need repair by breaking out a damaged area and refilling, such as crumbled areas, or tiny cracks which may need chipping prior to being filled in. Follow the manufacturer's directions carefully so that the new patch compound will adhere to the crack or holes and provide a secure filler.

Concrete substrates often necessitate the removal of old flooring when there are two or more layers. When one or more of those layers contain hazardous substances, removal should be done in accordance with accepted practices (see Chap. 7). All paints or other foreign elements must be removed from the surface of the concrete. Anything that would impede a satisfactory bond with the adhesive for the flooring should be taken off the subfloor. The volatility of floor covering adhesives requires you to be absolutely certain of the type and condition of the surface that will be gone over, and whether there exist any old flooring or adhesive residues. Chemical reactions between old *mastics* (adhesives) and new mastics can cause a bond failure. Consequently, be diligent in your search for clues as you investigate the premises.

Be as observant as possible so that you can evaluate the existing conditions and arrive at the answers to any possible questions. Write down all these observations, so that the shop drawings you provide to the installer will be as complete as possible.

Shop drawings

Shop drawings are the drawings provided by the floor covering contractor to the installer, and they are an indispensable tool in the successful completion of any project. They help determine and identify yardage, existing conditions, seam locations, transition moldings, wall moldings, and any other information pertinent to the job. This information may be included on a shop drawing:

- The name, address, and telephone number of the customer
- The date of the installation
- Drawings of all areas to have new flooring material
- Existing conditions of each substrate to be covered
- A list of solutions to any subfloor problems
- Location of all doorways and possible transition molding
- Type of flooring to be used in each area
- Type of installation method to be used for each section
- A list of all cuts of material, showing their length and where they are to be used on the diagram
- Excess flooring and how it will be utilized
- Direction of the pattern or pile
- Type of wall molding to be used, if any
- Location and type of stairs to be covered

A typical example of a shop drawing for a sheet vinyl flooring installation in a kitchen is shown in Fig. 1-19. It gives everyone a clear picture of the work ahead and enables the installer to come to the job with all the necessary tools and supplies. Another important feature about shop drawings is that they are useful in allowing the installer and the contractor to communicate intelligently in person or over the phone if there is a problem on the job. Since both people have a copy of the same diagram and information, it's easy to explain and solve a particular problem.

Significance of the adhesive
When you inspect locations where remodeling is to take place, you will find a variety of old mastics. Dealing with these adhesives will be easier when you understand the ingredients of modern-day products.

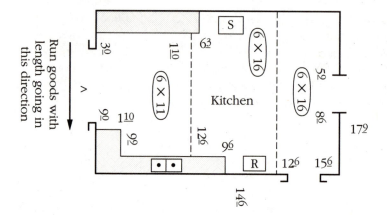

Instructions

Existing conditions - now - old asphalt tile on plywood with black cut-back adhesive. Leave in place. Do not remove.

- Nails heads popping from underneath the tile creating visible bumps, on the surface.
- Needs new 1/4" underlayment.
- R/R stove and refrigerator
- Needs 28 L/F - 4" rubber wall base on toe-kicks of the cabinets.
- 18" pattern-match on 6"-0" wide goods.
- Bring appropriate adhesive and seam sealer.

1-19 *Shop drawing.*

Many previously manufactured adhesives were of the solvent-emulsion type that contained a high ratio of solvent to water. Emission concerns from the volatile organic compounds (VOCs) found in floor covering adhesives prompted the flooring manufacturers, at the request of the U.S. Environmental Protection Agency (EPA), to voluntarily reduce emissions from their products. After extensive testing, products have been developed that provide low odor and low emissions of VOCs, for both resilient flooring and carpet adhesives. These ingredients are used in floor covering adhesives:

- *Binders*—a polymer, either natural or synthetic, that may consist of one or more resins. Binders produce the principal bond with the substrate.
- *Carrier*—an inactive liquid, generally water or petroleum by-products (solvents), needed to make the adhesive smooth and easy to spread.

- *Fillers*—used to add strength, an inert material typically made of calcium sulfate, calcium carbonate, clay, or sand.
- *Thickening agent*—additive that gives the adhesive its consistency.
- *Preservative*—chemical that protects the adhesive from fungus and micro-organisms.

The degree to which existing residual adhesives must be removed will depend largely upon whether the mastic is asphalt-based or latex-based and upon the type of adhesive to be used for the new flooring. Consult the appropriate floor covering or adhesive manufacturer before you proceed with any installation that has a questionable surface.

Blueprints and specifications

The combination of the architectural blueprint drawings and the accompanying specification details constitutes what is known as *a set of plans*. You will often be called upon to analyze floor covering material and installation costs for a structure that is yet to be built, and the only way to do that is to do a *takeoff* (cost analysis) from a set of plans.

The first time someone looks at a set of plans, it can be a mind-boggling experience. By breaking down the information into manageable components, however, the process of completing a takeoff becomes less complicated. The best way to break it down is to understand the functions of both the specifications and the drawings.

The specifications are the written instructions pertaining to a particular project. The blueprint drawings are the actual *scale* drawings of the structure itself. The specification package includes such information as bidding requirements, contract requirements, and products to be furnished. The architectural blueprint drawings show sketches of the floor plan, exterior elevations, plot plans, and detail drawings.

Specifications

Before you can begin the takeoff, you must thoroughly familiarize yourself with the specifications of the project. Most specifications will follow a format set forth by the *Construction Specifications Institute* (CSI) and *Construction Specifications Canada* (CSC). CSI and CSC formerly followed a specification guideline called the *Uniform Construction Index* (UCI). The UCI organized information into a coherent sequence that standardized all the various aspects required to complete a construction project. This 1972 document has been updated to its 1995 version and is now referred to as *MasterFormat*. The current edition is much more comprehensive and easy to understand. (Although a general overview of it is presented here, you would be well advised to obtain a copy for your own use.)

MasterFormat[1] is divided into six general headings that include the section title and any coordinating numbering system:

- Introductory information (numbered 00001 to 00099)
- Bidding requirements (00100 to 00499)
- Contracting requirements (00500 to 00999)
- Facilities and spaces (no numbering)
- Systems and assemblies (no numbering)
- Construction products and activities (divisions 1 to 16)

Introductory information shows such items as project title page, table of contents, and a list of drawings and of schedules. The bidding requirements section shows the instructions to the bidders, the information available to bidders, the bid forms, and bidding addenda. The contracting requirements section includes notice of award, agreement forms, bonds and certificates, as well as all the general and supplementary conditions. The facilities and spaces section identifies the requirements for complete constructed facilities. The systems and assemblies section provides information required to perform certain operations regarding such items as cost reports and performance requirements. The construction products and activities section is composed of 16 divisions which break down the construction process into understandable units of construction products and activities.

The 16 divisions of the construction products and activities are

Division 1—General requirements
Division 2—Site construction
Division 3—Concrete
Division 4—Masonry
Division 5—Metals
Division 6—Wood and plastics
Division 7—Thermal and moisture protection
Division 8—Doors and windows
Division 9—Finishes
Division 10—Specialties
Division 11—Equipment
Division 12—Furnishings

[1] MasterFormat was published by the Construction Specifications Institute in 1995.

Division 13—Special construction
Division 14—Conveying systems
Division 15—Mechanical
Division 16—Electrical

When you do a takeoff for a given set of plans, you must acquaint yourself with all the preliminary requirements. Once you have done so, however, it is the division 16 section of MasterFormat upon which you will concentrate most. Each division and resulting specialty code are based upon a five-digit numbering system. The first two numbers denote the division title (that is, 09 is division 9, finishes, while 12 is division 12, furnishings). The third number further categorizes products in a certain division [that is, 093(00) is tile and 096(00) is flooring]. The rightmost two numbers are even more specific (that is, 09310 is ceramic tile, 09330 is quarry tile, 09640 is wood flooring, and 09680 is carpet.)

Since you will be primarily concerned with division 9, it is essential to familiarize yourself with its contents. The information and trades covered are as follows:

Division 9—Finishes[2]

09050	Basic finish materials and methods
09100	Metal support assemblies
09200	Plaster and gypsum board
09300	Tile
09400	Terrazzo
09500	Ceilings
09600	Flooring
09700	Wall finishes
09800	Acoustical treatment
09900	Paints and coatings

If, e.g., you are going to concentrate solely on the flooring section, it is subdivided into the following groups. Here is a partial list of those subdivisions:

09630 **Masonry flooring**
Brick flooring
Chemical-resistant brick flooring
Flagstone flooring
Granite flooring
Marble flooring
Slate flooring
Stone flooring

[2] Reprinted by permission of the Construction Specifications Institute, 1996.

09640	**Wood flooring**
	Cushioned wood flooring assemblies
	Mastic-set wood flooring assemblies
	Resilient wood flooring assemblies
	Wood athletic flooring
	Wood block flooring
	Wood composition flooring
	Wood parquet flooring
	Wood strip flooring
09650	**Resilient flooring**
	Resilient base and accessories
	Resilient sheet flooring
	Resilient tile flooring
09680	**Carpet**
	Carpet cushion
	Carpet tile
	Indoor and outdoor carpet
	Sheet carpet

The size and scope of your operation will determine how many of the above products you will be able to bid on. If a particular set of plans calls for granite tile, wood parquet flooring, resilient sheet flooring, and sheet carpet, you might want to bid on all those different floor coverings. Often the various products will be listed as separate bids, so if you do not win the granite tile project, e.g., you might get the carpet or wood parquet flooring portion.

When you study the sections that you are interested in bidding on, you will notice that the exact type of material to be bid on is often specified. Often the manufacturer's name, the style name, and the color are stated. Certain descriptive information is also given regarding yarn quality, yarn weight, or other important facts about that particular product. However, if a precise product is not specified, certain requirements regarding minimum features are usually listed. Regardless of how it is stated, there will be some information listed that will enable you to arrive at a product suitable for the project. Once you have determined what that product is, you can calculate the cost after you have computed the amount of materials and the labor needed to complete the job. To compute the necessary materials, you must look at the blueprints.

Architectural blueprint drawings

A complete set of architectural blueprint drawings provides detailed sketches of the proposed structure. Accurate illustrations, from the floor plan to the elevations, make it possible for someone to view these drawings and get an exact interpretation of how this building will look once it is completed. The information on these pages should be so complete that little is left for the general contractor to decide. It includes all essential building elements such as plot plan, foundation plans, elevations, sections, and a number of clarification details.

To fit all this information on standard blueprint paper, which is roughly 24 in × 36¼ in, it must be drawn to a reduced scale. Depending upon the size and complexity of a project, certain portions of the drawings could have a scale equivalent of 1 in = 100 ft or 1 in = 1 ft. A typical residential floor plan, e.g., will usually have a scale in which ¼ in = 1 ft or ⅛ in = 1 ft. Each portion of the drawing will always show precisely the scale equivalent, so there is never any doubt as to what is understood. It will state, directly below the drawing, e.g., Scale ¼ in = 1 ft.

The dimensions of a given section of a scaled drawing can be converted to actual size by using either a scale ruler or a scale tape measure (Fig. 1-20). These tools allow you to instantly determine the precise length of any distance. You can then figure the length and width of a room to calculate the actual square footage, or square yardage, of that area. With that information, you can do a takeoff of

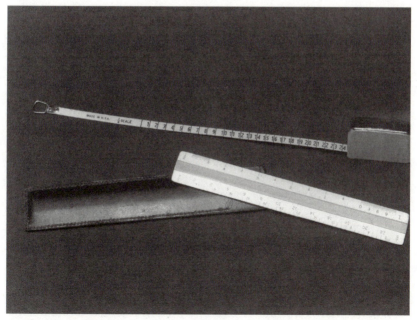

1-20 *Scale ruler and scale tape measure.* (Photo by Mahendra Ramawtar.)

any set of plans. Before you begin doing a takeoff, however, it is best to study the entire set of drawings so that you get an understanding and "feel" for the overall project.

Plot plan

A plot plan is a bird's-eye view of the entire piece of property. It shows the property boundaries, utility lines, any local *benchmark,* any existing structures, the proposed structure, certain trees, out-buildings, driveways, and the like. It gives you a basic orientation to the project. In addition, the compass designation pointing north and the surrounding street names enable you to pinpoint the location.

Floor plans

The floor plans are the portion of the drawings that you, as a flooring contractor, will concentrate on the most. They show in intricate detail (as if you were looking down on them from above) the entire layout of the structure. Since many of the jobs you might bid on will be homes, a single-family residence is used for this example.

A floor plan shows all the walls, partitions, room designations, dimensions, and a multitude of other significant details (Fig. 1-21). It also shows almost all the most important information regarding your proposal. Although you also need to examine the elevation drawings to see how the structure is designed, or to look at the detail drawings to inspect the floor construction, the floor plans remain your primary focus. Study them intently and read everything on them.

Room finish schedules

A *room finish schedule* is a columnar table that usually appears in a set of plans. It lists all the rooms and the various finishes for each (Fig. 1-22). For example, it explains what areas get flooring, wall base molding, paint, or other required finishes. Check the column entitled Floors, and note all the rooms to receive carpet, vinyl flooring, tile, wood, or any product you are going to bid on. Also check the Walls column to see whether there is any rubber wall base, carpet base, or tile base. Floor covering proposals often require the inclusion of specific wall moldings as part of the price. If the specifications or room finish schedule indicates that a product is to be included in the bid and you fail to do so, you will be obliged to provide and install those materials at your expense, if they accept your bid.

Elevations

While the floor plans merely indicate the length and width of a structure, the elevation drawings show the height. A typical elevation plan

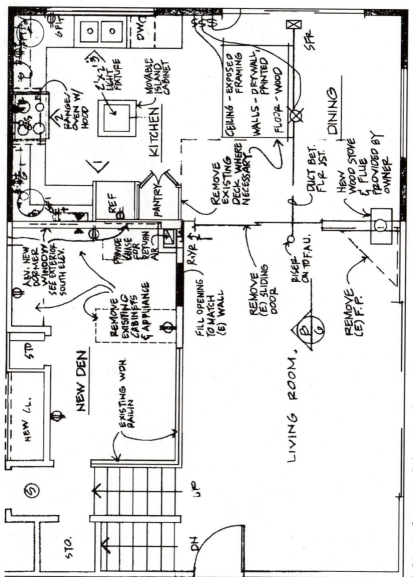

1-21 *Architectural drawing floor plan.* (Holehouse Construction.)

ROOM FINISH SCHEDULE

ROOM NAME	CLG HEIGHT	FLOOR MAT'L	CLG MAT'L	FINISH	WALL MAT'L	FINISH
ENTRY	8°	B	F	J	F	J
LIVING ROOM	8°	B	F	J	F	J
FAMILY	N/A	—	—	—	—	—
KITCHEN	8°	B	H	K	G & H	K
LAUNDRY NOOK	±9°	CONC.	G	K	G	K
BATH 3	PITCH	B	H	K	H	K
STAIRS	PITCH	A	F	J	F & G	J
BATH 1	8°	B	H	K	H	K
BATH 2	PITCH	B	H	K	H	K
MSTR BDRM.	PITCH	A	F	J	F	J
CLOSET	PITCH	A	F	J	F	J
BEDROOM 2	PITCH	A	F	J	F	J
BED 1	PITCH	A	F	J	F	J
BALCONY	N/A	STL.	—	—	—	—
UNDER STAIR STORAGE	N/A	B	G	K	G	K
GARAGE	±9°	CONC.	G	K	G	K

FLOOR MATERIAL
A-CARPET o/ PAD
B-SHEET VINYL
C-HARDWOOD
D-CERAMIC TILE
E-QUARRY TILE

CEILING MATERIAL
F-5/8" GWB
G-5/8" type 'X' GWB.
H-5/8" W/R GWB.

WALL MATERIAL
F-5/8" GWB
G-5/8" TYPE 'X' GWB.
H-5/8" W/R GWB.

FINISH NOTES
J-FLAT ACRYLIC VINYL o/ GWB
K-SEALED & PAINTED w/ TWO COATS-
  TOP COAT EGGSHELL ACRYLIC
L-

BASE
5 1/2" HIGH 711-TRIM

DOOR TRIM
BULLNOSE GWB.

1-22 *Room finish schedule.* (Peter Walker Hunt.)

illustrates all four sides of a structure. It presents the north elevation, the south elevation, the east elevation, and the west elevation. Looking at these elevation drawings, you will be able to visualize what the building will look like completed.

Elevations (also drawn to scale) show such elements as the sizes of the windows and doors, the pitch of the roof, the existence of any dormers, the natural grade of the surrounding terrain, chimneys, outside staircases, and a great deal more. Although you may not specifically utilize the intricate detail presented on the elevation drawings in formulating your floor covering proposal, you should familiarize yourself with the overall design of the building.

Sections

A section drawing is a view of a portion of the building as if part of it were cut away and the viewer were facing the remaining piece and looking at the inside configuration. A section line on a floor plan shows the imaginary cut, and it is usually indicated by a heavy double dot or dashed line. Indicators at either end of the dashed line point out the direction of the viewer's perspective. Also, within the indicator circle is the section drawing letter and page number (Fig. 1-23). Therefore, in this example, you know that this section drawing is *section C* and appears on page 6 of the working drawings.

Section drawings (here also to scale at $\frac{1}{4}$ in= 1 ft 0 in) are helpful in that they show the construction method and materials used for the floors, walls, stairs, and roof. They also provide valuable information on heights, angles, and a number of other fundamental elements. In addition, section drawings indicate and refer to detail drawings (Fig. 1-24).

Detail drawings

Detail drawings are blown-up portions of isolated areas that are critical to the strength and design of the structure. They are indicated by a circle on the elevation drawing, which specifies the area to be isolated, and an accompanying reference circle split in two, designating the detail number and the page where it will appear (Fig. 1-25).

The detail, in this instance, is important to you because it clearly shows the type of floor to be used in this area. You now know that the floor will be made of a single layer of plywood sheathing. Since

1-23
Section designation marker.
(Holehouse Construction.)

SECTION 'C'
1/4" = 1'-0"

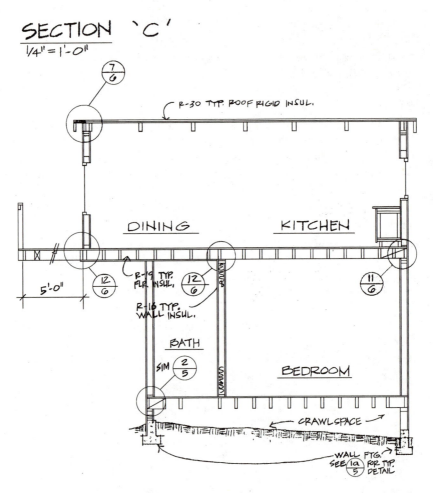

R-30 TYP. ROOF RIGID INSUL.

DINING   KITCHEN

5'-0"

R-19 TYP.
FLR. INSUL.

R-16 TYP.
WALL INSUL.

BATH

SIM

BEDROOM

CRAWLSPACE

WALL FTG.
SEE 1a FOR TYP.
5 DETAIL

1-24 *Detail section drawing C.* (Holehouse Construction.)

the scale for this drawing is 1½ in = 1 ft 0 in, you can also calculate that the plywood is resting on joists that are 16 in on center by using the scale ruler.

Architectural drawings and specifications give you all the necessary information to calculate a floor covering proposal. Once you have studied all the various sections pertinent to your trade, it is time to create the proposal.

Bidding on the job

Begin by separating the rooms on the plan to determine the floor materials for each. Remember, despite the complexity of the information presented to you in a set of plans, the actual working up of the proposal is no different from any ordinary project you bid on daily. If you debunk the myth of how complicated it is to figure a job from a set of plans, you will be more comfortable when you sit down to do one. Like so many other things in life, the more you do, the easier it gets. When you break the plans down into smaller units, this does simplify the process a great deal.

Start the takeoff by making all the appropriate measurements and computing the necessary yardage (or square footage) for the material needed. Then list all required products and compulsory extras together in logical form to arrive at the final price. For the floor plan in Fig. 1-21, the takeoff is prepared separately for the carpet and vinyl flooring.

The carpet is to be installed in the living room, in the dining room, and on the staircase. Those areas together require approximately 64 yd^2 of carpet. Since padding does not have the same restrictions as carpet (i.e., the lay of the nap and its availability in 12-ft widths), it is generally possible to use about 5 to 10 percent less padding than carpet, depending upon the actual sizes and layouts of the rooms. For our purposes, estimate the padding at 60 yd. In addition to those main items, there are sales tax, freight, installation labor costs, extra labor costs for specialty items like stairs or carpet coving, and any necessary preparation work. After you have assessed all those items, your takeoff should look like this (prices are at cost):

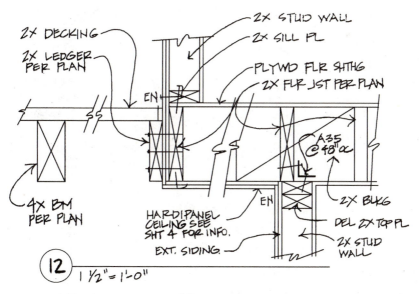

1-25 *Specific detail drawing* $\frac{12}{6}$. (Holehouse Construction.)

Carpet, 64 yd² @ $15/yd	$960.00
Padding, 60 yd² @ $2.95/yd	177.00
	$1137.00
California sales tax @ 7¾ percent	$88.12
Total material costs and sales tax	1225.12
Contract carpet installation @ $3.50/yd	224.00
Extra labor for 8 stairs @ $5.00 per stair	40.00
Freight charges for carpet delivery	32.00
Preparation work	35.00
	$1556.12
Profit and overhead @ 40 percent	622.44
Total	$2178.56

This is a fairly basic proposal, but it is one that you can use as the framework for most carpet installation bids. There are some variables on each job—the obvious ones are yardage, preparation work, and labor costs—but the structure of the proposal will be the same.

The sheet vinyl proposal is prepared just as the carpet proposal, except there will be additional materials to consider. The kitchen, in this example, requires approximately 12 yd² of sheet vinyl flooring. Also it needs an underlayment installed over the subfloor. Similar to the padding, the underlayment amount is less than the sheet vinyl amount because it covers less area. The amount of underlayment needed is 10⅔ yd² [or 96 square feet (ft²) which equals 3 sheets of 4-ft × 8-ft underlayment]. In addition, 48 lineal feet of 4-in top-set rubber wall molding will be necessary. Also you must add for the required floor covering adhesives, wall base adhesives, any metal edging strips for the transition between the carpet and the pad, as well as the compulsory items such as sales tax, freight, and labor charges. Consequently, the sheet vinyl proposal could look as follows (prices at cost):

Sheet vinyl floor covering, 12 yd² @ $19.00 yd	$228.00
Flooring adhesive, 1 unit @ $17.00	17.00
48-lineal-ft 4-in rubber top set @ $0.85 per lineal foot	40.80
Cove base adhesive, 1 unit @ $6.00	6.00
3 sheets (96 ft²) underlayment @ $16.26 per sheet	48.78
	$340.58

California sales tax, 7¾ percent	26.39
Contract vinyl installation 12 yd² @ $6.00/yd	72.00
Contract installation underlayment, 96 ft² @ $0.60/ft²	57.60
Contract installation 4-in top set, 48 @ $0.50 per lineal foot	24.00
Freight charges for vinyl delivery	20.00
Preparation work	35.00
	$575.57
Profit and overhead @ 40 percent	230.22
Total $805.79	

As you can see, the format is the same as that for the carpet proposal, only the components have changed. If you follow this simple formula, there aren't many jobs in the world that you couldn't bid on. There are, of course, other ways to figure markups and margins, so find the way that works best for you, and stick with it.

Estimating 1, 2, 3

The actual estimating of a job is as easy as 1, 2, 3:

1. List the price and all necessary materials and labor costs.
2. Calculate all sales taxes, freight charges, and incidental amounts.
3. Add the profit and overhead percentage.

Be diligent in your evaluation of every project, and make sure you cover all the required items. Leave out nothing. If you follow the three steps above, you can put all the pieces of the puzzle together. These steps can also be used for preparing bids for the use of any other materials, whether they be wood flooring, ceramic tile, or any floor covering product whatsoever.

2

Resilient flooring

Resilient, or hard-surface, flooring is a practical, durable product designed to combine superior wear resistance with contemporary styles. The extraordinary blends of patterns, textures, and colors are nearly limitless. Modern technologies have brought the construction and design of these products to a near art form. Every month or so, new products and designs are introduced to replace older ones. The industry is constantly evolving. It takes its cue from carpet manufacturers who are consistently experimenting with new color combinations. Therefore, resilient flooring is now more pleasing in appearance and is more durable now than at any time in the past.

The performance properties of resilient flooring are unsurpassed by other flooring products and are suited for both residential and commercial uses. According to the Resilient Floor Covering Institute (RFCI), until recently, the terms *linoleum* and *vinyl* were more commonly used to describe hard-surface flooring. Today, the term *resilient* is being used to depict more accurately the *attributes* of these products. The definition of *resilient* describes the materials' characteristics appropriately. "Resilient" means something capable of withstanding shock or compressive stress without suffering permanent deformation or rupture. This definition truly makes clear the intent of the design and construction of materials classified as resilient flooring. The ability to recover after the stress of constant traffic is the hallmark of this kind of product.

Types of resilient floor

A number of floor covering products fall into the category of resilient flooring:

- Linoleum
- Sheet vinyl flooring

- Vinyl composition tile (VCT)
- Solid vinyl tile
- Rubber tile
- Cork tile

Linoleum

Linoleum was one of the first resilient sheet floorings ever produced. While variations of this product are still being produced today, its use is not as widespread as it once was. Linoleum is a floor covering made by laying a mixture of solidified linseed oil, cork granules or wood flour, or a combination of these products, along with pigments of color on a burlap or canvas backing. (It is important to be aware of this product and its makeup because you will often encounter it when you are measuring older homes for new flooring.)

Linoleum is an interesting product. It has distinctive characteristics unlike those of contemporary sheet flooring. Because linoleum was made with linseed oil, if the flooring was left unattended and unwaxed for a long time, the linseed oil began to dry out and the floor became brittle and cracked. These cracks are called *crazing*. A product is said to craze when minute cracks appear on the surface. Another example of crazing is also evident on antique china or old ceramic tile. The enamel or glaze begins to show signs of fracturing, yet the surface is still smooth. In linoleum, however, the surface is not smooth. Its appearance is quite rough, and it has peaks and valleys caused by the dryness. This dryness and crazing often preclude installation of a new flooring directly on top of linoleum because the linoleum may require sanding to make it smooth, and sanding is strictly prohibited since the old linoleum may contain asbestos (see Chap. 7). *Never* sand a floor that might contain asbestos.

Technological advances after World War II brought about the development of sheet vinyl flooring. Sheet vinyl floor covering offers more diversity in pattern selection and color choice than linoleum ever did.

Sheet vinyl

Sheet vinyl flooring represents one of the largest segments of resilient floor covering products sold in the United States today. There are basically two types: rotogravure vinyl floors and through-chip inlaid products. Each product serves its own unique purpose in the way it benefits the modern consumer. Each has its own appearance and structural composition.

Rotogravure sheet vinyl

Rotogravure sheet vinyl flooring takes its name from a turn-of-the-century photographic process called *photogravure*. Photogravure is a process for making prints from an *intaglio plate* prepared by photographic methods. An intaglio plate is an engraving (or incised figure) in a hard material, depressed below the surface of that material, so that an impression from the design yields an image in relief. The term *rotogravure* describes a photogravure process in which an impression is produced by a rotary press.

With these definitions in mind, it is easy to see how the floor covering industry has kept the term *rotogravure* as part of its vernacular, because this type of floor covering is produced by printing a pattern on a base of vinyl and foam (Fig. 2-1). In addition, most of the rotogravure products have a pattern relief that evidences the gravure process. This relief not only demonstrates the manufacturing method, but also serves a functional purpose in that it helps to hide any imperfections in the subfloor upon which it is laid. If a floor covering were totally smooth, every nail head or subfloor joint would telegraph through the surface.

Because rotogravure vinyl floors are affixed to a layer of foam, they are more flexible in their construction and, therefore, easier to install. Also, since many of these products are thin and pliable, they are easier to seam together. Inasmuch as a great many come in both 6-ft- and 12-ft-wide sheets, ease of installation makes them ideal products for the do-it-yourself nonprofessional or the apprentice floor layer. While rotogravure vinyl floors are suitable for light to medium traffic, through-chip inlaid floors are advisable for heavy use.

Inlaid sheet vinyl

Through-chip inlaid sheet vinyl flooring is composed of small chips of solid vinyl granules, and it is one of the most durable sheet flooring products on the market today (Fig. 2-2). The color and pattern go all the way through the material. There is no foam backing as in rotogravure vinyl flooring, so the surface wear layer is less likely to be cut or damaged.

Inlaid sheet vinyl is made for residential as well as commercial applications. The residential products come with a no-wax surface for ease of maintenance, while most commercial inlaid products require a floor finish to help protect them and give them an appealing appearance.

Although both rotogravure and inlaid sheet vinyl can be *self-coved* (or *flash-coved*), (extending the material up the wall for 4 to 6 in), the inlaid sheet vinyl products are the most widely used for commercial

2-1 *Rotogravure sheet vinyl flooring. Congoleum Futura "Royal Gallery."* (Courtesy of Resilient Floor Covering Institute and Congoleum Corporation.)

coving purposes. Their durable nature makes them an excellent choice for restaurants, health care facilities, and any number of medium- to heavy-use industrial operations. Self-coving is used primarily to help keep bugs and other small, invertebrate animals from entering the premises under the wall molding. This is the reason why most local health departments require food service facilities to install sheet flooring that is coved up the walls in the food preparation and food storage areas. Otherwise, those areas could be contaminated by a legion of offensive vermin. Self-coving also prevents moisture from getting under the floor material. It also prevents dirt and grime from accumulating at the wall line because it eliminates the 90° angle between the floor and the wall. The gentle slope of the material going up the wall makes it easier to clean the floor. Even though self-coving can be used in a private setting for the same reasons, it is more often employed in residences for decorative purposes.

Since inlaid flooring is thicker and more difficult to work with than rotogravure products, it is best to have it professionally installed. The typical do-it-yourselfer would not have, or know how to use properly, the necessary tools. Apart from that, it takes years of experience to learn how to install these products correctly. Align yourself with a good sheet vinyl installer so the jobs you contract can be installed by a seasoned journeyman.

Vinyl composition tile

Vinyl composition tile (VCT) is made of vinyl chips and fillers that are well integrated throughout the entire thickness of each tile (Fig. 2-3). This mixture is spread out in a doughlike fashion and then is cut into separate tiles. Each tile is made to an exact thickness and is cut to precise, clean measurement specifications so that the tiles fit together perfectly. Since the vinyl chips go all the way through the product, the surface design will never wear off completely. Regardless of the amount of traffic it might receive, the original veining characteristics will remain.

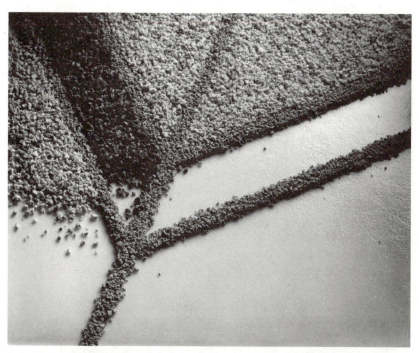

2-2 *Through-chip inlaid sheet vinyl flooring manufacturing process from vinyl granules.* (Courtesy of Armstrong World Industries, Inc., Lancaster, PA.)

2-2 *Continued.*

2-2 *Continued.*

2-3 *Vinyl composition tile. Azrock commercial flooring "Classic Granite" with feature strip.* (Courtesy of Resilient Floor Covering Institute and Azrock Industries.)

Prior to the introduction of vinyl composition tiles in the 1970s, this type of tile was composed of, and called, *vinyl asbestos tile* (VAT). When it was discovered that asbestos was a leading cause of some cancers, production of VAT was halted and vinyl composition tile was introduced in its place. Over the years, there have been relatively few complaints regarding the qualitative differences between VAT and VCT. Vinyl composition tile has proved to be a long-wearing, dependable product that can be used under extremely rigorous conditions.

It is important to note that there is a tremendous amount of vinyl asbestos tile installed in both residential and commercial structures. Removal or installation of new flooring over VAT must be carefully handled. (See Chap. 7.)

Vinyl composition tiles come in commercial and residential styles. The commercial products are produced as described above and require a floor finish or wax as part of a regular maintenance program. Many residential tiles are not necessarily through-chip products, but they do come with a clear polymeric surface, no-wax coating that adds shine to the tile. (Each manufacturer has its own process and name for the coating, so consult them for more specific details.) This coating is extremely helpful in facilitating upkeep, because dirt cannot penetrate deeply into the surface of the tile.

Residential vinyl composition tile also comes with either dry backings or self-stick backings. The dry-backing tiles require the use of an adhesive, while the self-stick types have had an adhesive applied at the factory. Self-stick tiles are perfect for homeowners who wish to install the floor themselves. Self-stick tiles come in hundreds of designs and colors, from a number of different manufacturers. These products offer an inexpensive to medium-priced option for covering a floor.

Solid vinyl tile

Pure solid vinyl tiles have a much denser structure and are slightly more resilient than vinyl composition tiles. Solid vinyl tile is manufactured through a *compression molding* or *nondirectional chip* process (Fig. 2-4). This process entails heating and molding pure vinyl chips, filler, and color pigments in a press to make the product more supple. This process also increases the design capabilities. Inasmuch as solid vinyl is a molded and heated product, the color is evenly distributed throughout the tile and therefore does not wear away. Also, its dense structure makes it highly resistant to stains and indentations.

In the past few years, solid vinyl tiles have gone through significant design changes. More patterns are being developed that replicate hardwood, stone, or ceramic tile. Despite the fact that it is a premium-priced product (one of the most expensive of the resilient floors), consumers have responded to it because it offers the flexibility of a custom design coupled with superior performance qualities.

Rubber flooring

Smooth, flat surface-molded rubber floor tile has been on the market for many years. It has been used quite successfully in high-traffic

2-4 *Solid vinyl tile. Congoleum "Classic Touch Colosseum."* (Courtesy of
Resilient Floor Covering Institute and Congoleum Corporation.)

areas. Its pattern design creates a marbleized effect that gives it the
same appearance as a vinyl composition tile. The rubber tile that is
most widely used today, however, consists of a pattern of raised circular disks or squares. Rubber floor tiles (Fig. 2-5) are made from
high-quality, properly cured, homogeneous rubber compounds.
Therefore the color extends throughout the thickness of the tile.
Rubber tiles contain no asbestos fiber and have a sanded backing
to ensure proper adhesion. The size of the tiles and their thickness

2-5 *Raised rubber floor tiles. Burke "Rouleau" rubber tile and stair tread.* (Courtesy of Resilient Floor Covering Institute and Burke Industries.)

depend on the individual manufacturer. (Consult the appropriate producer for specific details.)

The unique raised design is ideal for heavy-traffic areas such as lobbies, aisles, and ramps. These products also come in premolded stair treads. The raised design provides traction and prevents slippage that could be caused by a slick, wet surface. Dirt and water do not accumulate on the surface; the moisture falls between the raised portions of the tiles, so the top portion remains dry. As a result of this excellent traction feature (even when wet), rubber flooring is being specified more frequently by architects and designers in areas where pedestrian slippage could be a problem.

Since rubber flooring has an extremely long life span—perhaps 15 to 20 years—it is excellent for educational and health care facilities as well as commercial and industrial areas, such as malls and airports. Residential use of rubber flooring, incidentally, is on the rise. It offers a clean, contemporary look to most rooms, and many decorators are specifying it for use in kitchens, family rooms, and game rooms.

Cork flooring

Cork flooring is a natural and environmentally friendly flooring system that is seeing a resurgence in use today. The early cork floors came sanded and waxed, whereas many of the newer products, such as those from Ipocork Company, come with a polyvinyl chloride (PVC) backing and a transparent PVC top laminate (Fig. 2-6). The raw material—cork—comes from the outer skin of the cork oak tree. Every 9 years, after being stripped, this species has the ability to grow a new bark (Fig. 2-7). When the tree dehydrates in the hot summer months, the bark is loosened and the skin is removed and placed in piles. The drying process then takes about 3 months. After that it is boiled in water mixed with fungicides for increased strength and pliability. Finally, it is left in a dark cellar for 1 month before it is ready for processing into floor tiles. After the bark has been removed, the tree will grow another complete layer of bark and be ready for another harvest almost a decade later.

Cork tiles offer a variety of features quite different from human-made material. Cork is a natural wood product so each cork tile is unique in color and grain pattern. Millions of tiny air cells produce a natural cushion that gives the floor a noticeable flexibility. This cushion allows the tiles to recover from temporary indentations. The natural thermal and acoustical insulation characteristics of cork keep the floor warm in the winter and cool in the summer. In addition,

1. LMT — Low maintenance treatment
2. Tough transparent wear layer
3. Natural or colored cork veneer
4. Resilient agglomerated cork
5. Embossed synthetic backing

2-6 *Cross-section of a cork floor tile.* (Copyright Amorim Group—Portugal.)

2-7 *Harvesting bark from a cork tree.* (Copyright Amorim Group— Portugal.)

because cork has such excellent acoustical properties, the tiles reduce noise levels in the rooms where they are installed.

From an installation standpoint, cork must be treated the same as any wood flooring product, so certain precautions should be taken. A $\frac{1}{8}$- to $\frac{1}{4}$-in expansion space around the perimeter of the room must be left void because temperature and humidity changes will cause the floor to expand and contract; without the expansion gap, the floor will eventually buckle. The tiles should also be allowed some time to acclimate to the on-site environmental conditions. Similar to hardwood flooring, it is best to let cork sit on the job for a minimum of 24 h at temperatures of at least 68°F before installation.

Cork flooring is perhaps one of the truest renewable raw materials used to manufacture floor coverings in the world today. Unlike hardwood flooring, where the trees are cut down as they are harvested, cork oak trees are not destroyed in the process of gathering their yield. The cork trees remain intact, and the bark simply renews itself. New plantations of cork oaks guarantee production for many years to come.

Since cork flooring is visually so attractive and extraordinarily practical, its uses are suited to both commercial applications and private residences. A vast array of decorative designs and colors give cork tiles a unique niche in the floor covering market of the future.

Although it is a high- to premium-priced product, its versatility in both form and function should not be overlooked. It makes a stunningly beautiful floor.

Installation of tile flooring

The procedure for laying out a floor to install tile is the same regardless of the type of tile being used. Whether the tile is vinyl composition, ceramic, or parquet, arranging the layout is the same.

Properly laying out a room before spreading any adhesive is the first step toward a good installation. Two primary pattern formations can be used to lay tile: the square-to-the-wall or the diagonal-to-the-wall. In either instance, ensure that the tiles laid around the perimeter of the room are no less than half a tile. Tiles that are cut less than that size look unsightly and unprofessional. So, to achieve that appearance, certain adjustments may need to be made when you make the preliminary calculations.

The best layouts are designed to have the tile installed from the center of the room outward, because not all rooms have walls and floors that are square to each other. (Although a room might roughly measure 12 ft × 12 ft, in some places there could be a variance of substantial size.) The center of the room is established by measuring the length and width of the room, regardless of any variance in straightness such as setbacks, jogs, or recesses. Simply measure the true length and the true width. Determine the halfway point of each end wall, mark that point, and then use a chalk line to establish a straight line between the two walls. Striking a line on the floor separates the room into two sections along the length (Fig. 2-8).

Determine the center point of line *wy/xz* by accurately measuring it and marking the position. This point becomes point *a*. From point *a*, on that line measure an equal interval on each side of it. The distance can be any amount, but typically it is less than 6 ft. These spots become points *b* and *c*. Then, choosing a measurement that is greater than the distance between *a* and *b*, scribe two arcs that are perpendicular to line *wy/xz* from points *b* and *c*. A chalk line is then marked through the arcs that go through point *a* at right angles to line *wy/xz* (Fig. 2-9). The room is now divided into four equal sections. To ascertain the exact border size, take accurate measurements to ensure that it will be at least half of a tile. If it is smaller than half of a tile, adjust line *wx/yz* and/or line *wy/xz*.

To lay out a room for tile to be laid on a diagonal, begin by squaring the room as shown above. Along line *wx/yz*, from point *a*,

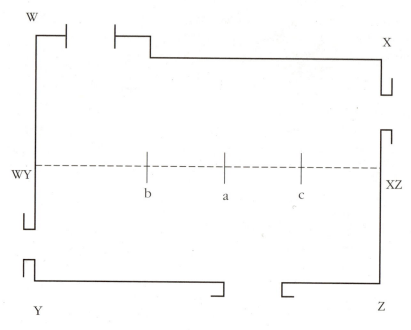

2-8 *Sketch showing how to separate a room into two sections.*

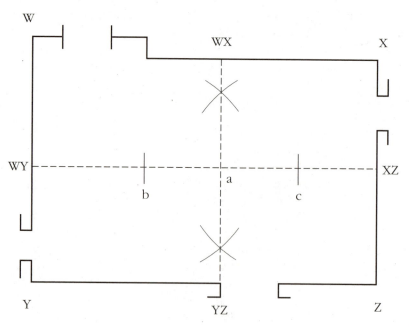

2-9 *Sketch showing a room divided into four quarters.*

establish two points in the same manner as points *b* and *c* were determined. These become points *d* and *e*. Using a measurement that is greater than *a* to *b*, scribe an arc from *b*, *c*, *d*, and *e*. This will form line *wz/xy*, which will be on a diagonal to *wy/xz* and *wx/yz* (Fig. 2-10). If the border tiles are not at least half of a tile, adjust diagonal lines *xy* and/or *wz*.

Since all manufacturers have certain specifications for the adhesives that can be used with their products, consult with them to be sure that the mastic chosen is appropriate. Once that issue has been resolved and the adhesive has been spread, it is time to lay the tile. (Be careful, when you spread the adhesive, to come right up to the chalk line. Do not cover the chalk line. Spread the adhesive evenly so there are no pools or empty areas.)

The first tile that is placed in position is the most critical tile. This tile will determine the success of the entire installation. Gently place it at the intersection of the chalk lines. This tile must be placed exactly on the lines. Then press firmly on the tile to properly seat it. Next, working outward from the first tile and both guidelines, begin laying the tiles. The edges of the tile should be on the guidelines and set firmly to the adjoining tile. Do not slide the tile into position. Begin laying subsequent tiles in an alternating pyramid design after the

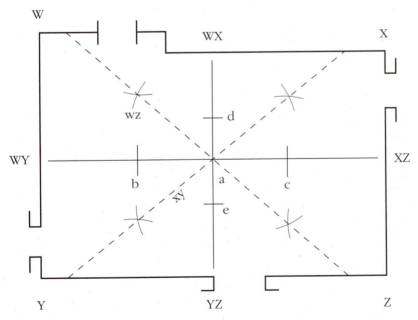

2-10 *Laying out a room to have tile laid on a diagonal.*

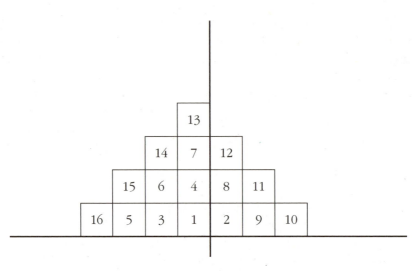

2-11 *Sequential order for laying floor tile.*

base row is started (Fig. 2-11). Continue this installation pattern throughout the entire area except for the border tile.

To cut the border tile in a square-to-the-wall installation, place a tile directly on top of the last full field tile. Then place another tile on top of that tile, but flush to the wall. Score the first tile, using the edge of the second tile. Cut the first tile and fit it into place. Use a stiff cardboard template for the diagonal border tile. The factory edge of a tile is cut very precisely and therefore should be placed next to the field tile, and the cut edge should be facing the wall. It is advisable to pattern or direct scribe the border tile when protrusions, pipes, or other irregularities exist in a room. A perfect fit will be achieved by scribing onto the tile to allow for any unusual circumstances. The resulting appearance will be professional and neat.

When you lay tiles on a diagonal, strike all square lines and all diagonal lines as shown in Fig. 2-10. Dry-lay tile along the diagonal lines to ensure proper tile alignment and border width. Any adjustments can be made at this point by moving the diagonal lines accordingly. If the chalk lines are correctly struck, you may need to start with four tiles on the centerline, instead of one tile directly in the middle of all the intersecting lines. Once you feel comfortable with how the tiles look when laid out dry, spread the adhesive and lay the tile along the diagonal centerline. Depending upon how you want the room to look, either you can lay border tile square to the wall, to give a picture-frame appearance, or you can continue the tile all the way around in a diagonal fashion, with half tiles around the perimeter and quarter tiles in the four corners.

The installation of floor tiles offers limitless design possibilities. The use of such patterns as checkerboards, or any number of multicolored tile formations, is bound only by the designer's imagination. In addition, damaged tiles are easily repaired. Simply remove one piece and replace it with another. Sheet vinyl flooring, however, is much more difficult to repair and to install.

Sheet vinyl installations

A great many products are available from the major sheet vinyl flooring manufacturers, and each producer has its own *specific* installation recommendations. Since this book does not explain the installation procedures for each individual product, contact the appropriate manufacturer to obtain the applicable installation instructions before you proceed with any job. However, many factors common to almost all sheet vinyl installations are worth examining.

All manufacturers have their own accessory products to accompany their flooring materials. They have their own adhesives, seam sealers, waxes, and the like. These products are typically not interchangeable from one manufacturer to another. The use of another sheet vinyl manufacturer's products could void the warranty of the original manufacturer. There are, however, separate manufacturers who specialize in accessory products such as adhesives or seam sealers. If products are used from these suppliers, it is the responsibility of the accessory product manufacturer to warrant the performance of its own product. For example, if a certain adhesive is specified by the flooring manufacturer, the installer elects to use another product, and a bond failure results, then the sheet vinyl company—assuming the installation is done properly—is not liable for that problem. The installer has to contact the accessory manufacturer to file a claim. If that proves unsuccessful, the installer is responsible for any damages.

Since sheet vinyl manufacturers have spent a lot of time and energy perfecting products that enhance the performance of their materials, usually they are only willing to guarantee their own products. This is not to say that there are not quality independent products on the market that are equally good. It is just best to be aware of potential problems given a product failure. Balance the savings against the risk; then make a decision. The type of flooring, and particularly the type of backing on the sheet goods, determines the interchangeable nature of any adhesives. The more specialized the backing, the more likely it is you should stay with the recommended adhesive. Generic-type felt backings can often be installed with a premium-grade adhesive.

Installation systems

There are two basic methods of installing sheet vinyl flooring: *full-spread* and *perimeter* installation. These two terms refer to how, and where, the adhesive is applied to the floor covering material.

In the full-spread method, the adhesive is spread underneath a sheet vinyl flooring that has a nonasbestos, felt backing. This type of installation represents the traditional approach to installing sheet flooring. An adhesive is spread throughout the entire area and secures the sheet goods firmly to the substrate. With the perimeter method, the adhesive is spread only in a 3- to 4-in band around the perimeter of the room and at the seams. The center portion of the room is left without any adhesive. The backing of these sheet goods is a vinyl backing, not to a felt backing.

In the perimeter method, after sheet goods are cut and installed, they have a tendency to *shrink* or to develop *tension*. This shrinking helps the material lie flatter, and it also allows any air bubbles created during the installation to find an escape route from under the flooring. These air bubbles may take 72 h or more to disappear, so be patient. Alert your clients to this possibility, and instruct them not to put heavy objects, such as books, on top of the bubbles to speed up their dissipation. The bubbles will disappear on their own. Another good feature of this installation method is that it has a tendency to hide minor imperfections in a subfloor. These sheet goods have more *give* and will flex more over surface irregularities.

In contrast, full-spread installations require more precise preparation procedures because the sheet goods will be drawn more tightly into the subfloor as the adhesive begins to set up. If there is a flaw in the substrate, it may eventually telegraph through the sheet goods. Therefore, be sure to prepare the subfloor properly before you lay any resilient flooring product..

General conditions

Regardless of the installation method chosen, the substrate must be clean, smooth, dry, and free of wax, paint, oil, grease, varnish, solvents, or any foreign matter that could create an adhesive bond failure.

The room temperature for sheet vinyl installation should be between 65 and 80°F. This temperature should be maintained for 24 h prior to the installation and for at least 48 h after the job is completed. Also, allow both the flooring products and adhesives to come to room temperature prior to installation. This is particularly important in the colder regions of the country. During the winter, the materials should be brought to the site at least one day before installation and put in the room where they are to be installed.

All resilient sheet vinyl flooring must be rolled with the patterned surface facing outward. If sheet flooring is rolled with the pattern facing inward, the goods have a tendency to curl at the edges. Also, it will make the goods harder to secure around the perimeter of the room, and it will create problems in seaming the two pieces together.

Whenever possible, sheet vinyl flooring should be left rolled and standing on end. When rolls of this material are laid down horizontally for any length of time, the bottom side of the roll tends to flatten. Subsequently, when the goods are unrolled, there are waves in the material that may or may not flatten out. Seaming products in that condition is difficult, if not impossible.

Embossing levelers

Whenever a new resilient floor is to be installed over an existing, single layer of residential sheet vinyl flooring or tile that has an embossed wear surface, instead of removing the floor, an *embossing leveler* may be used. An embossing leveler uses a mixture of portland cement and a special liquid latex that is mixed on the job and troweled onto the existing sheet vinyl flooring. It fills and levels the embossed texture of the original flooring prior to the installation of the new materials. If this is not done, the old pattern could telegraph through the new flooring.

An embossing leveler may be used on an existing floor regardless of whether the new floor will be laid down with the full-spread method or the perimeter installation method. It is important that any old wax or floor finishes be stripped from the surface of the existing flooring before you apply the embossing leveler; otherwise the flooring may not stick. The leveler must dry for about 2 h before you install the new flooring. Of course, temperature and humidity conditions may lengthen or shorten that drying time. Since all major floor covering manufacturers have their own embossing levelers, contact them for the specifications and use instructions for individual products.

Pattern matching

The seaming together of two or more sheets of vinyl floor covering is the most critical feature in the success of any installation. Since most sheet vinyls have a distinct pattern, it is important to match the pattern accurately. In floor covering, similar to wallpaper, the design must be aligned correctly, or else the seam will be glaringly obvious. To eliminate this possibility, each manufacturer has unique techniques for addressing this issue. However, some basic principles can be applied for most products.

Design match runout

One of the most common problems in sheet vinyl products is pattern stretching. The pattern and wear layer can stretch when the material is bent around tubes as the sheet goods are made into rolls. The amount of stretch will vary in proportion to the diameter of the tube, whether the individual cut of material was from the center of the roll or the outer edge of the roll, and how tightly the roll was wrapped. These three factors greatly influence the actual installation of the flooring in terms of layout and design match.

Keep in mind that the stress on a small-diameter tube is considerably greater than that on a large-diameter tube. (Because of the structural differences from product to product, the recovery factor will vary as well. Hard, stiff-backed sheet goods will recover differently from soft, vinyl-backed material.) The tension around a 4-in tube will be greater than that on a 6-in tube. This will result in more pattern stretch on a roll with a 4-in tube.

The sections of the flooring cut from the outer edges of the roll have less strain on the material than cuts made closer to the tube. In addition, the tightness of the wrap may vary from roll to roll, resulting in design differences in rolls of the same pattern, in the same sequential run. The actual stretch difference from the center to the outer portion of a roll could be $\frac{1}{2}$ in or more. Consequently, it is imperative to install the pieces in the exact sequence in which they are cut from the roll. Mark the cuts 1, 2, 3, and so on (in pencil, not felt-tip pen, because inks may bleed through the backing into the surface) to keep the sequence well documented. If cut 3 were installed next to cut 1, the pattern stretch could make it difficult to align the design properly.

As a rule of thumb, it is best to lay out the room so that the shortest possible seams are used. Thus the likelihood of design runoff is greatly reduced.

When a design is not lining up properly, it may be possible to manipulate the sheet goods to bring the pattern *onto match*. If the pattern is stretched more than its neighboring piece (because it is tightly rolled on a tube), temporarily reroll it with the backing facing outward to loosen the tension. Also, if possible, precut all lengths of material a day before the job is to be installed, and reroll each cut into a separate roll of the same diameter to achieve uniform tension.

When you match a pattern, begin the match in the center of the seam line. By doing so, any variance is equally divided to either side of center, lessening the visual impact of a mismatched pattern. Also, make actual job site measurements of the pattern before you make any cuts from the roll, to ensure true matches.

For installations that require more than one roll of material, the sequence of cuts should be reversed. For example, if the first roll was cut from the outer edges to the tube and was marked 1, 2, 3, then the second roll should be cut in the same order but marked for laying as 3, 2, 1. This means that the piece which was cut last is installed first. As a result, the piece that was closest to the tube on the second roll is next to the piece that was closest to the tube on the first roll. There-fore, there is a greater likelihood that the stretch of these two adjoin-ing pieces will be similar and will produce closely approximating pattern repeats. Conversely, if the outer edge of the second roll were installed next to the inner cut of the first roll, the stretch could be vastly different, making a good pattern match very difficult.

Consult the manufacturer's instructions to determine if individual sheets of material should be laid in the reverse-sheet method (turned 180° in the opposite direction) or in the do-not-reverse method (same direction).

Seam cutting

The three following methods can be used to make seams: straight-edge and butt, recess-scribed, and double-cut.

Double-cut seams are produced by overlapping two pieces of sheet vinyl flooring and cutting through both layers at once (Fig. 2-12). The seams should be cut before any adhesive is placed directly under them. A scrap of excess sheet goods should be laid upside down un-der the two layers of flooring. This helps protect the tip of the blade. But it also adds fullness to the cut so that the edges will be a slightly separated. Then when the adhesive is spread under the seam, the seam can be hand-rolled in place (Fig. 2-13). When that is completed, the entire job should be rolled with a 75- to 100-pound (lb) floor roller, if called for by the manufacturer. Any seam sealers should be applied at this time.

Recess-scribed seams

While double-cutting is designed for use on products that are thin or soft and can have two layers cut through easily at the same time, ma-terials that are thick and hard require an alternative method. *Recess scribing* is a procedure employing a specialty tool called an *under-scriber* (Fig. 2-14). An underscriber is a mechanism that scribes a line with a hard metal needle on a piece of sheet goods which slightly overlaps another piece of sheet goods. It is a two-pronged apparatus whose bottom portion has a knob that follows along the edge of the

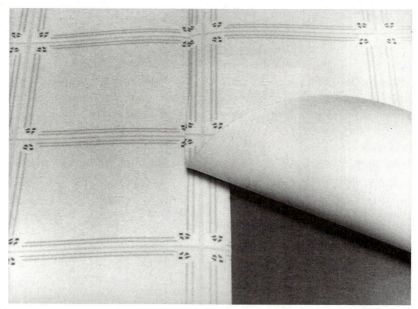

2-12 *Photo showing how to line up pattern to double-cut a seam.* (Photo by Mahendra Ramawtar.)

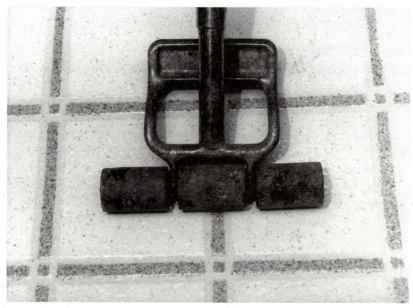

2-13 *Use of a hand roller to roll in a seam.* (Photo by Mahendra Ramawtar.)

piece underneath. (This bottom layer should be trimmed first with a straightedge to remove the factory edge.) The top portion of this apparatus has a sharp needle that is set directly above the seam line of the piece on the bottom. As this device is moved down the trimmed bottom seam, it scratches a score line in the top layer that matches exactly the line underneath. The installer then makes a clean cut on the score line of the top piece to complete the trimming of the seam. When the adhesive is in place underneath the seam, the installer can gently press the two pieces together. They should fit together perfectly. The seam is then hand-rolled into place. After that the entire job is rolled with a 75- to 100-lb roller to *seat* all the adhesive.

Straightedge and butt seams

Only certain products should have their seams straightedged and butted together. To do so, straightedge the first sheet of material. Then place that piece over the second sheet. After you ensure that the pattern match is correct, slide the metal straightedge up against the first seam that has been straightedged. Lift the trimmed piece away, and the metal straightedge will be in position to trim the bottom piece accurately. Make a clean cut, and the pieces should properly fit together. Coat seams with any appropriate seam sealers to finish the job.

2-14 *Sheet vinyl seam scribing tool called an underscriber.* (Photo by Mahendra Ramawtar.)

No matter what installation method you use, flooring adhesive should not be allowed to enter the seam when two pieces of material are placed together. When this happens, it is called *seam contamination*. The adhesive that contaminates a seam can darken or otherwise highlight the seam, making the seam more visible. It can also interfere with the chemical structure of any seam coatings and can cause further problems.

This section on resilient flooring reviewed how these floors are manufactured and installed. As you become more familiar with these products, you will realize how vastly different they are, yet how similar their job site requirements are. Because you are dealing with a thin material and an adhesive, precise preparation work is critical. Any imperfection in the subfloor can potentially compromise an otherwise superb installation. Therefore, thoroughly check out everything in advance, and take the time to do it right. You save time in the long run.

3

Carpet

Historically, the use of carpet (or, more appropriately stated, rugs) dates back several thousand years. The Persians, Greeks, Chinese, and Romans all had rug weaving as an important element in their cultures. Much desired for their warmth and beauty, early rugs were usually the exclusive possessions of royalty. Exquisite rugs made of silk and gold covered the floors and walls of many a castle. Rug weaving, done by hand, was a long and laborious task.

As time went on and the demand by the greater populace increased, the need to improve production became obvious. By the early 18th century, two rug weaving centers were established in England, one in the town of Wilton and the other in the town of Axminster. (To this day, the names *Wilton* and *Axminster* survive, and they have become synonymous with two particular ways to manufacture carpet. These methods are discussed later in this chapter.)

In the 19th century, carpet was manufactured in the United States on mechanical looms that produced a fabric 27 inches wide. These narrow carpets were sewn together by hand, until there was sufficient width to cover the floor of a room. Gradually, the quality of the machines and of the carpets improved until the U.S. producers were making a product similar to the woven carpets from Axminster and Wilton.

In the 1920s, the countryside surrounding Dalton, Georgia, was filled with quilters and bedspread makers. This was a cottage industry in which entire families produced bedspreads and quilts in their homes. Travelers going through the area during that time reported to have seen bedspreads hanging on the clothesline of every cabin, drying in the sun after the products were washed. The people who produced these bedspreads and quilts used a customized sewing machine that created a tufted design in the fabric. The idea of manufacturing carpet by tufting fibers onto a backing material was developed by individuals aware of this unique bedspread industry.

The first carpet tufting machines were actually converted bedspread tufting machines. They began by fabricating carpet that was still the basic 27 in wide. Many improvements were made over the years, and by the late 1940s producers were able to manufacture tremendous amounts of carpeting in their factories. The tufting machines had been redesigned to produce carpets in several wider versions: 9, 12, and 15 ft. These wider carpets came to be known as *broadloom* carpet, relating to an obviously wider loom than the old 27-in standard. The term *broadloom* is still used today to refer to wall-to-wall carpeting. Although the most common width of carpet used today is 12 ft, some manufacturers make 15-ft-wide carpeting and more recently carpeting that is 13 ft 6 in wide. The latter width is used primarily for Berber carpets so that seaming can be minimized.

Over the years, carpet manufacturing has grown into a multibillion-dollar industry that is constantly striving to improve its product. Currently, high-speed, computerized tufting machines are producing state-of-the-art quality products. Carpets are available for purchase to fit any budget or satisfy any special design requirement. To this day, Dalton, Georgia, remains the center of the carpet industry in the United States.

Elements of carpeting

Carpeting is not just a single item; it is an entire system. That system consists of the carpet itself, the pad, and the installation. Carpet also creates and defines the ambiance of a room. It is a visual cue, and it is an emotional sensor. It provides warmth and comfort and it has color, style, and texture.

Traditionally, carpet is chosen to cover the floors of more rooms in a given building or home than any other floor covering product. On a square-foot basis, it is one of the least expensive floor coverings to install. Yet it seems as if it is the most expensive because typically it covers so much more area than, say, sheet vinyl or ceramic tile.

To choose the correct components for any carpet system, you must understand some basic principles and know the characteristics of the products involved:

- *Fiber.* A carpet fiber is a filament which comes from a variety of raw materials. The source of the threads greatly affects the carpet's durability and stain-resistant properties.
- *Construction.* There are several ways to manufacture carpet, and each one serves a unique purpose.

- *Style.* Although there are only a few basic styles, color and fiber variations make it appear as if there are limitless numbers.
- *Color.* Color is what makes carpet such an emotionally influential element.
- *Texture.* The numerous textures available help to establish the *use* guidelines for carpeting (i.e., formal or casual).
- *Cushion.* The carpet cushion that is installed underneath a carpet greatly affects how a carpet will perform in a given situation.
- *Installation.* The installation method (whether stretched in or glued down) typically is determined by the job site conditions and the general requirements of the owner.
- *Maintenance.* The kind of care a carpet needs is an influential factor in a customer's choice. A high-maintenance carpet may often be passed over for one that requires less maintenance.
- *Warranties.* Accompanying warranties are of extreme importance to contemporary consumers. Performance assurance is essential to buyers purchasing something so important for the home or business.

All the above attributes and concerns strongly influence a potential client who is beginning to shop for carpeting. Knowing how to address those concerns and elaborate on the attributes of the items you recommend will be very much to your benefit when you explain the virtues of any carpet system.

Carpet fibers

Carpet fibers are divided into two basic categories: natural and artificial. The natural fibers consist of wool, coir, sisal, cotton, and jute. Artificial, or human-made, fibers include nylon, polyester, olefin (polypropylene), and acrylic. In any carpet purchase, the fiber content should be the primary consideration. The most commonly used fibers are the following.

Nylon

It has been reported that the term *nylon* originated during World War II. At that time, the Allies were attempting to develop a product to replace silk for the manufacture of parachutes. The United States and England were collaborating on the project, and the primary work was being conducted in New York City and London. So by taking *ny* from New York and *lon* from London, they coined the word *nylon*. Once produced, nylon proved so versatile that many other uses were immediately found. One of the more successful applications of nylon was its use in women's stockings. Nylon stockings soon replaced silk stockings because nylons were less expensive and possessed incredible strength. As the years passed, the use of nylon increased, and it ultimately began to be used for the manufacturing of wall-to-wall carpeting.

Today, nylon is approximately 67 percent of the carpet fiber used in the United States. One of its most important features is its *memory*. Memory is a measure of a fiber's ability to bounce back, as from foot traffic. Fibers that lack memory have a greater tendency to quickly mat down and do not have the capacity to regain their original appearance. Another word for memory is *resilience*. Just as with hard-surface flooring, resilience refers to a product's ability to withstand, and recover from compressive stress.

Nylon also has outstanding performance capabilities for soil resistance, stain resistance, appearance retention (matting and crushing), and wear resistance. These qualities, coupled with the fact that nylon is affordably priced, make it the premier choice for carpet fiber. With an endless variety of styles, colors, and textures, nylon fibers dominate the carpet industry at present.

Under the nylon umbrella, there are *branded fibers*. These are fibers that have specific names, from specific yarn manufacturers. They are known as *advanced-generation* fibers because they possess properties for stain resistance and wear resistance that were unheard of before their inception in the early 1980s. With these fibers the consumer is given express, written warranties from the fiber manufacturer (which supersede most carpet manufacturer warranties) that cover the performance of the fiber itself. The introduction of these carpet fibers has been a boon to the carpet industry.

Fabrication of nylon fibers

Nylon is produced into two basic yarn forms: *bulked continuous filament (BCF)* and *staple yarn*. Nylon, a petroleum-based product, is made into long, thin strands that can measure up to 12 ft or more. These strands are called *continuous-filament yarns*. When tufted onto a backing material in this long, continuous state, the fiber has a propensity to snag and unravel when pulled. A Berber carpet is a good example of a continuous-filament yarn manufactured in continuous strands. BCF yarn, however, can be produced for a cut-pile carpet by clipping the top loop of the yarn. One of the best features of a BCF cut-pile or a continuous-yarn system is that the fiber does not shed. Thus carpet has a cleaner finish and always looks neat, even between vacuum cleanings.

Nylon yarns that do shed fibers once they are manufactured are called *staple yarns*. Staple fibers are produced by cutting a piece of BCF into smaller strands that measure from 3 to 9 inches long. Once the BCF is cut to create staple fibers, the yarn begins to shed because the continuous strand has been compromised. This shedding will continue indefinitely, but it does tend to slow down slightly after about 8 to 10 vacuum cleanings. Staple yarns are more voluminous, and consequently they can be made into carpets that appear heavier than they actually are. It will take a greater amount of BCF yarn to manufacture a carpet that is equal in appearance to a carpet manufactured with staple fiber yarn.

After the filament yarns are produced, they are *heat-set* by a method that uses a combination of pressure, steam, and heat to permanently *set* the twist in the yarn so that it has greater resistance to matting and crushing. This process gives the yarn its resilience and bounce.

Olefin (also called polypropylene)

Olefin fiber is a synthetic material that is strong, durable, and resistant to permanent staining. It is also resistant to moisture and mildew and is therefore ideal for use in carpets referred to as indoor/outdoor carpets. In addition, it is resistant to static electricity. Since the color is added during the fiber production process (solution-dyed fiber), olefin produces a colorfast fiber that is less likely to fade due to sun exposure or bleach spillage.

Inasmuch as olefin is one of the fibers that lacks the memory of a nylon fiber, it is less resilient and has a tendency to mat. Since its cost is slightly lower, it is well suited for carpets oriented to a budget-conscious consumer. Many residential Berber carpets and tight commercial carpets are constructed with olefin fibers. Every year more and more carpets are being produced with olefin fiber. It is the second fiber of choice, next to nylon.

Polyester

Use of polyester fiber results in a carpet that is exceptionally soft and luxuriant. In thick, dense cut-pile carpets, it has a *hand* (feel) that is rich and lush. Since polyester is also a somewhat less costly fiber to produce than nylon, carpets can be manufactured that have more bulk and fiber content per square yard at a lower price. Polyester, like olefin, lacks the memory of nylon and has a tendency to mat down.

Polyester has superb color clarity and produces vivid carpet colors that are often difficult to reproduce in a nylon carpet. Polyester fibers are resistant to water-soluble stains and are therefore easy to clean.

Acrylic

Acrylic fibers are used most often today as a blend with wool fibers. Acrylic has the look and feel of wool, but at a much lower cost. Blends of wool and acrylic, however, do not have the performance capabilities of an all-wool carpet. Acrylics are moisture- and mildew-resistant and have low static electricity levels.

Wool

Wool fiber, a product shorn from sheep, is the most widely used natural fiber for carpeting. It is flame-resistant, water-repellent, and very durable. Wool fibers have a *natural memory,* inherent in the structure of the fiber, which allows it to retain its shape. This unique, natural resilience is formed by a spiraling crimp that lets the wool stretch and compress much as a spring mechanism does (Fig. 3-1).

Wool is the most flame-resistant of all carpet fibers. Its high moisture content and protein constituents will not support combustion (Fig. 3-2). While synthetic fibers will melt, when exposed to burning embers, wool will merely char.

Wool has a natural ability to absorb color and retain it for the life of the fabric. Color is drawn to the center of the shaft, where a molecular bond forms that makes wool amazingly colorfast (Fig. 3-3).

Wool has an ability to hold dirt high in the pile so that it can be released from the fiber when vacuum-cleaned. Overlapping scales on the shaft of the fiber keep dirt from penetrating deep into its core (Fig. 3-4).

3-1
Wool fiber's spiral crimping provides superior resilience.

3-2
Wool fibers have a moisture content that makes them flame-resistant.

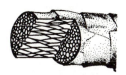

3-3
Wool fibers create a lifelong bond with color dyes.

3-4
Scales on wool fibers limit the penetration of dirt and dust.

3-5
Protective membranes on wool fibers allow them to actually shed water.

Wool is extremely water-repellent (Fig. 3-5). A thin protective membrane covers the fiber, thus providing a surface that sheds water. Also, because of its high moisture content, wool reduces static electricity and shocks.

Although wool carpet is generally thought of as a premium-priced product, there are now available lower-cost, excellent-quality carpets made from 100 percent wool fibers. Since wool possesses so many superb characteristics, it is worth considering as a carpet fiber option.

Coir

Coir (pronounced "koyer") is a stiff, coarse fiber made from the outer husks of the coconut. If you have ever held a coconut in your hands, you know how rough it feels. Imagine how it must feel on bare feet— but that is part of its appeal. It is a departure from the softer fibers so often used, and it gives a room a rustic, country-cottage look. Although coir is usually thought of as the fiber used in those thick, hairy doormats, it is also used for larger area rugs and broadloom (wall-to-wall) carpets.

Jute

Jute is a glossy fiber of either of two East Indian plants (*Corchorus olitorius* and *Corchorus capsularis*) used chiefly for sacking, burlap, twine, carpet padding, and carpet fiber. Like coir, jute is used for both rugs and broadloom carpet. In rugs, jute is often combined with cotton to produce a soft, uncommon feel and appearance.

Cotton

Cotton (*Gossypium*) is a soft fibrous substance thought of mostly for use in clothing, yet it is also used for flooring products. It is manufactured into area rugs more often than broadloom carpet. As a broadloom carpet, it crushes easily and is difficult to clean. Therefore, it is not a good choice for wall-to-wall carpeting.

Sisal

Sisal (pronounced "si-sell," "si-zell," or "sis-ell") is a strong, durable white fiber widely cultivated from the West Indian agave plant (*Agave sisalana*) whose leaves yield the fibrous product. Sisal plants have also been hybridized in East Africa.

Of the sisal, jute, and coir fibers, sisal is the most popular and most expensive. Sisal is used extensively for both broadloom carpet and area rugs. Sisal, like coir and jute fibers, is very difficult to keep clean. Stains from wine and some dark-colored fruit juices will be almost impossible to remove. A professional cleaner most likely will be required. Also, since these natural fibrous rugs absorb and give off moisture, a slight odor may be noticeable from time to time.

Despite the many drawbacks of jute, coir, and sisal carpets and rugs, they are on the cutting edge of decorating ideas for designers and architects alike because the look they provide is very adaptable. In one setting, these carpets will look homey and provincial, yet, in another location they will look high-technology and modern. These rugs and carpets also make excellent backdrops for oriental or contemporary area rugs. Any of these fibrous products when installed as wall-to-wall carpet with an area rug laid over it creates a stunning visual effect.

Carpet construction

The methods used for constructing carpets are: hand-knotted, woven, tufted, and needle punched. A vast majority (over 90 percent) of the broadloom carpet manufactured in the United States today is tufted. Woven carpets, which include Axminsters and Wiltons, are more costly to produce and therefore are used less frequently. Needle-punching is used primarily for specialty items and indoor/outdoor carpeting. Hand knotting is a woven process utilized strictly for handmade rugs.

Hand-knotted rugs

All rug making begins with a loom. One basic loom is the vertical loom. It is composed of two vertical poles secured in the ground (or permanently affixed in place), with horizontal beams strapped to them forming a roughly rectangular frame (Fig. 3-6). Vertical threads, called the *warps* (usually of wool, cotton, or silk), are wrapped tightly to the top and bottom poles. This establishes the substructure onto which yarn can be tied. The warps are placed side by side to form the entire width of carpet.

The weaver begins by creating several base rows of yarn at the bottom end of the loom. This flat-weave foundation, called *kilim*

3-6 *Rug loom.* (Photo courtesy of Durango Trading Company.)

(which means "no nap"), binds the warp threads together and provides the basis of the rug. The colored yarns that form the pattern and content of the rug (typically of wool or silk) can now be woven into place. Weaving and tying the yarn under one warp thread and over the next one, on a horizontal plane, create the *weft* threads. The artisan will continue weaving the pattern to the very top until the design is complete. Then another several kilim rows are added to lock the final edge of the rug in place and to protect it from unraveling. The rug is removed from the loom by cutting the warp threads. These threads become the fringe and are an integral part of the rug. The finished piece is then sheared, trimmed, cleaned, and set in the sun to dry. This general description of how rugs are woven by hand may be useful to know since many broadloom carpets have similar construction methods, even though they are produced on high-speed machines.

Woven carpets

Woven carpets are similar to handmade rugs in that the surface pile is firmly interwoven with weft fiber, around warp threads, along with backing yarn to create a single integrated textile product (Fig. 3-7). Woven carpets come in two types: Wilton and Axminster. The main difference between these two types is that Axminster carpets are limited to multicolored patterns in cut pile only, whereas Wilton carpets can create a similar carpet in cut-loop, variable-loop, and hard-twist pile, with or without patterns.

Unlike tufted carpets that have a secondary backing, woven carpets will not delaminate. Also, woven carpets are less likely to pull or snag because the *tuft bind* (the ability of a yarn to remain attached to the backing when pulled) is extremely strong (Fig. 3-8).

Woven carpets are constructed so securely and so densely that their production costs are very high. This makes a woven carpet a premium-priced product. Nevertheless, due to its incredible wear characteristics, it is recommended for heavy-use facilities such as hotel lobbies, casinos, and movie theaters. The wide array of colors and patterns that can be employed make them ideal for areas where there is a possibility of food or beverage spills. Consequently, the cost is far outweighed by its superior wearability and design versatility.

Tufted carpets

Tufted carpets are manufactured on high-speed tufting machines that sew large loops of yarn through a primary backing material. A layer

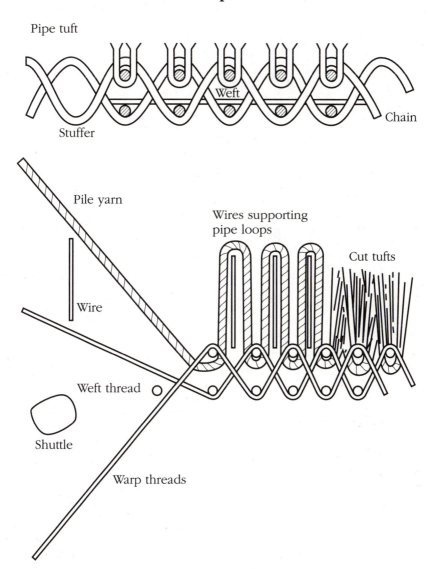

Pipe tuft

Weft

Stuffer

Chain

Pile yarn

Wires supporting
pipe loops

Cut tufts

Wire

Weft thread

Shuttle

Warp threads

3-7 *Construction method of a Wilton carpet. Top: A section of a piece of Wilton carpet. Bottom: A diagram illustrating how the carpet is constructed.*

of soft latex is then applied to the backing to help bond the yarn in place. A secondary backing is then attached to that latex coating to give the carpet its dimensional stability (Fig. 3-9). Tufted carpets are the most economical to produce and, therefore, account for most of the carpet manufactured today.

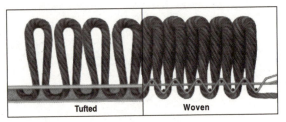

3-8 *Tufted versus woven carpet construction.*
(Reproduced with express permission of Crossley Carpet Mills Ltd.,
Truro, Nova Scotia, Canada.)

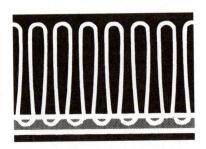

3-9
*Cross section of tufted carpet
construction.* (Photo courtesy of the
DuPont Corporation.)

Needle punching

Carpets produced using the needle-punch process are fabricated from
barbed needles that punch carpet fibers through a mesh fabric core.
The result is that needle-punch carpets are more difficult to unravel.
They are punched into place and not tufted. This locked-in process is
well suited for outdoor use where yarn pulls could pose a problem.

Carpet styles and textures

Choosing the correct carpet for a certain setting depends on both the
carpet construction and the style of carpet. A Berber carpet, e.g., may
look terrific in one environment, but could look dreadful in another.
Careful attention needs to be paid to the surroundings and furnishings of a certain area to arrive at the right combination of style, texture, and color.

Several tufted carpet styles are manufactured today. The *style* refers
to the appearance of the fabric in its finished condition—the look of a
carpet. The basic styles are Saxony plush, textured plush, level loop,
cut and loop, frieze (pronounced "friz-ay"), cable, multilevel level
loop, and random shear.

It is appropriate to group these styles into three main categories: cut pile, loop pile, and cut and loop. The classification of the styles is as follows:

1. Cut pile
   a. Saxony plush
   b. Textured Saxony
   c. Frieze
   d. Cable yarns

2. Loop piles
   a. Level loop
   b. Multilevel loop

3. Cut-and-loop
   a. Sculptured
   b. Random shear

Cut-pile carpet

Cut-pile carpet is commonly thought of as "traditional" carpet. It is the style of carpet that was once used most frequently. Cut-pile carpets have an even-looking surface because the tops of the loops are cut to produce tufts of yarn that stand in an upright position (Fig. 3-10). One excellent feature of cut-pile carpet is that, unlike loop pile construction, the yarn will not unravel because the top of the loop has been sheared, leaving individual tufts. From a visual perspective, two tops equal one strand, and that strand is not connected to the next strand. Therefore, the extent of any unraveling is limited to just the one strand of yarn.

All cut piles are constructed in a similar fashion; it is the texture of the yarn that changes the look and feel of the carpet.

3-10 *Cut-pile carpet.* (Courtesy of the DuPont Company.)

Saxony

Smooth Saxony plush carpet is often referred to as *velvet* or *velour*. It is very densely packed to provide a smooth, luxurious feel. This style of carpet is well suited for formal settings. Since the surface pile is so flat, this type of carpet is likely to show footprints and vacuum marks more than some of the other styles (Fig. 3-11).

Textured Saxony

The construction of a textured Saxony is similar to that of a Saxony plush except that the individual strands of yarn are twisted in such a way that the surface pile is not as smooth. As a result, the light refraction, as it hits the yarn, gives the surface more of a two-tone color effect. Although the yarn is one color, it appears as if there are specks of white (or lighter shades of the predominant color) within the fiber, but this is merely an optical illusion. Also, because of the twist factor, a textured Saxony will conceal footprints better than a Saxony plush (Fig. 3-12).

3-11 *Saxony plush cut-pile carpet.* (Courtesy of the DuPont Company.)

3-12 *Textured plush carpet.* (Courtesy of the DuPont Company.)

Frieze

The yarn twist of a frieze carpet is the tightest of all cut-pile carpets. The yarn is twisted to such a high degree that it appears to curl upon itself. It is then heat-set at a very high temperature so that the twist level remains. This process not only helps hide footprints, but also enables the fiber to resist crushing and matting better than almost any other type of yarn construction. The intensity of the twist acts as a spring which allows the yarn to bounce back more readily. From a decorating standpoint, frieze carpets are used more often to create an informal atmosphere (Fig. 3-13).

Cable yarns

Composed of larger-diameter tufts of yarn, cable yarn carpets have gained wider acceptance in recent years. Manufactured in a looser cut-pile method, cable yarns are reminiscent of shag carpets, but are much more elegant. They are soft underfoot and create an informal, yet tasteful ambiance.

Loop pile carpet

Loop pile carpet represents a large segment of carpet currently produced. It is manufactured by tufting long strands of continuous-filament yarn. Loop pile carpet includes Berber, patterned Berber, level loop carpet, and multilevel loop carpet.

Berber

The word *Berber* comes from the Berber tribes in northern Africa, particularly of Morocco. The style of their rug weaving, which gives a fabric its nubby texture, has become synonymous with contemporary Berber carpet (Fig. 3-14). Since the word *Berber* has become so broad in its scope of reference, as it relates to modern-day carpets, it encompasses level loop, multilevel loop, and even patterned loop types. One distinction, however, is that the term *Berber* typically refers only to residential carpet, not commercial carpet.

Patterned Berber

This carpet can have either a distinctive geometric multilevel loop or a random multilevel loop appearance in monochromatic or multi-color yarns (Fig. 3-15).

Level loop

These carpets are constructed of very tight, densely packed loop pile yarns of equal height. As with all types of loop pile carpets, since this

3-13 *Frieze carpet.* (Courtesy of the DuPont Company.)

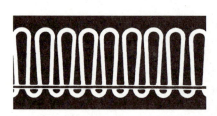
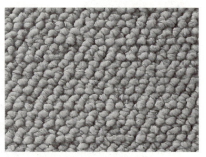

3-14 *Berber carpet.* (Courtesy of the DuPont Company.)

3-15 *Patterned Berber carpet.* (Courtesy of the DuPont Company.)

method utilizes an unbroken thread of yarn, if the fiber is pulled, it will unravel. Although this style of carpet offers excellent durability, it does suffer from that one, significant problem. Proper care, by cleaning with a suction-type vacuum cleaner, as opposed to a vacuum with a beater bar or rotary brush, is extremely important. A loose thread can catch on a rotary brush and pull out an entire row of yarn.

Once pulled, that thread can never be repaired or replaced. However, level loop carpets are good for high-traffic areas, because they have exceptional wear qualities.

Multilevel loop
Generally, this carpet has two or three different loop heights that give it a unique patterned effect. It is similar to the level loop in durability, but has a style all its own.

Cut-and-loop carpet

Cut-and-loop carpet contains yarn that is a combination of both cut and looped fibers. This provides a difference in surface textures that enhances the appearance of the carpet. The two types of cut-and-loop carpets are *sculptured* and *random shear.*

Sculptured
Used mostly for residential applications, sculptured carpet usually has high cut-pile tufts along with low loop tufts that are set farther below the top surface. It usually comes in multicolored tones that are great for hiding dirt and spills (Fig. 3-16).

Random shear
This carpet style is a mixture of cut and uncut loops that are at a more consistent height than those in the sculptured carpet. It creates a very textured appearance that is unusual and attractive (Fig. 3-17).

Carpet backing

Tufted carpets are constructed using two distinctive backing systems: the primary back and the secondary back. Each system plays

3-16 *Sculptured carpet.* (Courtesy of the DuPont Company.)

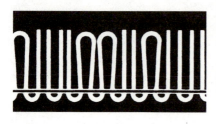

3-17 *Random shear carpet.* (Courtesy of the DuPont Company.)

an important role in the quality of the performance of broadloom carpet. The primary backing system is of one type of material, whereas secondary backings can be any of several materials. Jute, polypropylene, unitary, and foam are the most common secondary backing materials.

Primary backing

The primary backing is a tightly woven fabric (usually made of polypropylene) onto which yarn is tufted. When the yarn is tufted to the primary backing, the carpet is very pliable and does not have a strong tuft bind. It would be totally unsuitable for the forceful methods used in the installation process, such as power stretching and knee-kicking. Therefore, a secondary backing is required to withstand these rigorous installation practices.

Secondary backing

The secondary backing is attached to the primary backing with a latex adhesive to provide greater dimensional stability. This secondary backing stiffens the carpet and improves its performance capabilities. However, the most important function of a secondary backing is that it helps anchor the pile yarn to the primary backing to increase the tuft bind. The application of the secondary backing also curtails *edge ravel* (the tendency of yarn to fray at the side of the roll or at a seam). Secondary backings are made of the following materials.

Jute

Fibers of the jute plant provided the basis of the main secondary backing material used in the United States until the late 1970s and early 1980s. Because of importation problems from their source overseas, manufacturers found it necessary to search for alternative raw materials to produce secondary backings. Jute is an excellent backing material because it is strong and satisfactorily absorbs the latex adhesive. It seams well and holds a stretch adequately, but it has a tendency to mildew in damp environments because it is a natural fiber that absorbs moisture.

Jute is not as widely used today as it once was. Jute backings were replaced by polypropylene backings so that carpet manufacturers could have a constant supply of raw materials at a price that could be controlled.

Polypropylene

This type of secondary backing is not as closely stitched as is a primary backing. The threads appear as a perfect geometric pattern comprised of $\frac{1}{8}$-in squares. Like jute, it is attached to the primary backing by latex adhesive. It is by far the most widely used backing today. Besides being manufactured in the United States, another of its main advantages over jute is that it resists mildew because it does not absorb moisture.

Polypropylene is the least expensive of all backing materials. It is used primarily for stretch-in installations over padding, but it can also be used for glue-down installations. The problem that arises from gluing it down, however, is that the tuft bind is not very strong. This lack of tuft bind could lead to severe edge or even field ravel. It is best to use a unitary backing whenever possible for glue-down installations.

Unitary backing

This type of backing is not an actual fabric like a jute or polypropylene backing. It is merely a latex coating applied directly to the backstitch of the carpet. This coating tremendously improves the tuft bind characteristic of the yarn. It provides a 20-lb tuft bind that translates to a superior-quality carpet edge. (A 20-lb tuft bind is the ability of the yarn to stay attached to the backing, without raveling, if a tuft is hooked and 20 lb of pressure is pulling on it.) Unitary backings are designed only for direct glue-down installations. Special adhesives are required, and the carpet should be rolled with a 75- to 100-lb roller after it is installed.

Foam backing

Foam-backed carpets are made with an attached urethane cushion that is bonded to the back of the carpet. It is used in place of a secondary backing on tufted carpets, and it is intended to be glued down only. They do alleviate the necessity of an additional padding, while at the same time providing a soft cushioning underfoot. In addition to the advantages of a direct glue-down installation (i.e., no loss-of-stretch problems that create wrinkles in the carpet), foam-backed carpets have excellent performance characteristics. These products once had severe manufacturing problems, but they now have improved protection from *delamination* (when the primary and secondary backings separate), edge ravel, and moisture absorption.

Carpet dyeing

More than all other aspects of a carpet (i.e., style, texture, and fiber), it is color that creates the mood in a room. Color is dynamic, and it has an emotional impact on our senses. In most cases, the color of the carpet, in a home or office, is the first decorating feature that is noticed.

There are three primary colors—red, yellow, and blue—from which all other color possibilities are created. There are also three secondary colors—green, violet, and orange. Then there are intermediate colors between the primary and secondary colors. Color, in all its variations, comes from these basic combinations.The color we see is actually the light reflected "off" an object. The reflection of these colors can have a profound effect on how we emotionally perceive a setting. To create a feeling of warmth or to encourage activity, choose a warm color such as a shade of red, orange, or yellow. For a more introspective or calming effect, select a color from the cool palettes of blue, green, or violet. It is important to understand the major methods of color application for carpeting today so you can better assist clients in selecting carpets to decorate their surroundings.

Synthetic fibers represent the majority of fibers currently in use. These fibers come in round, trilobal, pentagonal, octagonal, hollow, and many other shapes. Each of these distinctive shapes will accept the color (dye stuffs) differently.

There are two possible times during which yarn can be dyed in the manufacturing process:

- Predyeing—the fiber or yarn is dyed prior to being tufted.
- Postdyeing—the carpet is dyed after it is tufted onto the primary backing, but before the secondary backing is applied.

As can be concluded by these definitions, predyeing colors the fiber itself whereas postdyeing colors the actual tufted carpet. The method used for a given product depends upon the effect desired by the carpet designer.

There are several methods for each system, and each has unique features. For example, one of the predyeing methods is called *solu-*

tion dyeing. In solution dyeing, the color pigment is added to the molten polymer before the extrusion process turns the polymer into fiber. This creates a fiber that is extremely colorfast and fade-resistant because the color is an integral part of the fiber. Solution-dyed products are less likely to be damaged by bleach or sunlight. The fibers most often used for solution dyeing are polypropylene and nylon.

In contrast, in the postdyeing method, there is a process called *beck dyeing.* In this procedure, *greige goods* (undyed, unfinished carpet) are moved in and out of a dye bath to allow for superior color uniformity. However, because the color is merely "applied" to the fiber, it is more likely to fade than a solution-dyed carpet.

These are just two examples of perhaps 15 or more different ways to dye yarn or carpet. Each method affects the way the yarn is colored and its appearance in the finished product.

Whatever the dye method chosen, the visual appeal of the carpet is the most important consideration. Carpet color is the foundation upon which most interior design is established. Choosing the correct carpet color creates a pleasing effect that will last for years.

Carpet cushion

Carpet cushion (also known as *padding*) is often thought of as necessary only to provide a soft feeling under the carpet. A bounce, or springiness, is what most consumers equate with the true value of a carpet cushion. This assumption is not necessarily true in all cases. It is often best to have a firm, dense pad that provides better stability and support for a carpet.

Carpet cushion serves a number of functions. It provides thermal insulation, noise control, increased vacuum-cleaning efficiency, and extended life of the carpet. In addition, it provides comfort, durability, and *appearance retention* (the ability of a carpet to look new for a long time, i.e., resistance to crushing and matting).

Carpet cushion falls into the following three basic categories:

1. Urethane foam
2. Fiber
3. Rubber

Within these categories, there are various types of products from which to choose (Fig. 3-18). Figure 3-18 shows not only the type of padding available, but also the minimum recommendations suggested by the Carpet Cushion Council for cushion used in different areas of the home. These recommendations have been adopted by

TYPE	KEY CHARACTERISTICS					
	CLASS 1			CLASS 2		
	weight, oz./sq. yd. min.	density, lbs./cu. ft. min.	thickness, in. min.	weight, oz./sq. yd. min.	density, lbs./cu. ft. min.	thickness, in. min.
URETHANE						
Prime		2.2	0.375	Not Recommended for class 2		
Grafted Prime		2.7	0.250		2.7	0.250
Densified Prime		2.2	0.313		2.7	0.250
Bonded		5.0	0.375		6.5	0.375
Mechanically Frothed		10.0	0.250		12.0	0.250
FIBER						
Rubberized Hair Jute	40.0	12.3	0.2700	50.0	11.1	0.375
Rubberized Jute	32.0	8.5	0.3125	40.0	8.9	0.375
Synthetic Fibers	22.0	6.5	0.2500	28.0	6.5	0.300
Resinated Recycled Textile Fiber	24.0	7.3	0.2500	30.0	7.3	0.300
RUBBER						
Flat Rubber	56.0	18.0	0.220	64.0	21.0	0.220
Rippled Rubber	48.0	14.0	0.285	64.0	16.0	0.330

Maximum thickness for any product is 0.5 inches.

Class 1: Light and Moderate Traffic such as living, dining, bedrooms, and recreational rooms.

Class 2 cushion may be used in Class 1 applications.

Class 2: Heavy Duty Traffic, such as lobbies and corridors in multi-family facilities, and all stair applications.

This chart illustrates products that meet **minimum** guidelines. Better grades of carpet cushion than the minimum suggested are always recommended when possible to provide more support and cushion for carpet. In areas where heavy use is expected, the Carpet Cushion Council suggests using firmer grades of cushion. These areas include stairways, halls, and areas where heavy furniture is used (such as living rooms and dining rooms). Softer cushion may be used in bedrooms and lounge areas where use is lighter and a plusher "feel" is desired.

*HUD requirements have been in use for over 20 years, and have been proven to work. In addition, the Carpet Cushion Council in cooperation with other associations, has conducted extensive testing on different types of cushion to see how they perform in use. Ask your floorcovering dealer for cushion that meets or exceeds these requirements. The UM72a standards include other technical characteristics for each type of carpet cushion. Please contact the Carpet Cushion Council for further details.

3-18 *Carpet cushion types and characteristics.* (Courtesy of Carpet Cushion Council.)

the U.S. Department of Housing and Urban Development (HUD) for use in FHA-financed housing. It should be stressed that these are *minimum* guidelines and that by upgrading the quality of the pad from these guidelines, the performance of the carpet will be increased as well.

Certain terms are commonly used regarding carpet cushion. A few of these terms are as follows:

Density This is the weight of the pad per cubic foot. For example, if a pad is described as a ½-in, 6-lb pad, it is ½ in thick with a weight factor of 6 lb per cubic foot (lb/ft^3).

K factor The rate at which a material conducts heat (thermal conductivity) is the K factor. High-K-factor materials are heat conductors, whereas low-K-factor materials are considered insulators.

R factor The insulating ability of a material is measured by its R factor. To arrive at the R factor, it is necessary to divide the thickness of the material by the K factor.

$$R \text{ factor} = \frac{\text{thickness, inches}}{K \text{ factor}}$$

Therefore, insulators have high R values, and conductors have low R values.

Compression force deflection (CFD)   This is the force necessary to compress 1 square inch (in^2) of padding thickness by a percentage of its original thickness. This is measured in pounds per square inch (psi).

Noise reduction coefficient   A carpet installed with separate cushion, as opposed to a carpet that is glued directly to the floor with no padding, provides greater sound reduction because the padding absorbs a lot of surface noise.

These are but a few of the terms pertaining to carpet padding, but they provide a basis for understanding its qualities. Since the different types of cushion have distinctive properties, they cannot be compared according to one test or feature alone. The construction of a particular product, and the intended use, will determine its suitability for a given installation.

The three primary categories of cushion—urethane, fiber, and rubber—have unique properties and characteristics. A look at these groupings will help you in evaluating a client's cushion requirements.

Urethane

Urethane carpet cushions come in prime urethane, bonded urethane, and mechanically frothed urethane. Prime urethane comes in three types: conventional, grafted, and densified. Conventional and grafted urethanes are produced by using a chemical mixing reaction process, whereas in the densified prime urethane the chemical structure is modified to achieve certain desired features. Prime urethane is manufactured from polyether urethane.

Bonded urethane foam is made of polyether foam as well as other products that sometimes include polyester urethane. These are shredded and combined through a fusion process to become one solid membrane, much as chips of wood are fused together to become particleboard. Mechanically frothed pad is a urethane foam applied to a sheet of unwoven fabric. It creates a strong, dense, durable pad for commercial or residential use.

Fiber pad

Fiber cushion products are typically made of natural or synthetic materials. The natural fiber products can be made of jute or animal hair.

The synthetic cushions are manufactured of polyester, polypropylene, nylon, or recycled textile fibers.

These cushions are generally used under Berber carpets or commercial carpets. Since commercial and Berber carpets are typically short, dense products, a cushion that is firm and does not have a lot of bounce is preferable because it prevents seams from *peaking* (the seam of a carpet protrudes upward and becomes prominent due to excessive movement of the padding underneath).

Jute and animal hair padding are used less frequently today because synthetic fiber pads have captured the market. Just as jute backing had to be phased out due to price and product availability problems in the late 1970s, the use of jute padding needed to be curtailed as well. Inasmuch as jute fiber tends to mildew in damp conditions, that was a contributing factor to its loss of popularity. Today, since synthetic fiber cushion is readily available and competitively priced and does not mildew, it is the pad of choice over jute and animal hair pads.

Rubber

Rubber carpet cushion is made of natural rubber, synthetic rubber, or a combination of both. It comes in a flat or rippled surface. The rippled sponge rubber pad offers a soft feel, while the flat rubber is much firmer.

The most important factor to consider in choosing a padding is that it is meant to work in conjunction—as a unit—with the carpeting. If the cushion is of an inferior quality, the useful life of the carpet will be compromised. Conversely, the correct cushion choice will provide years of favorable service. Note that many fiber manufacturers and carpet mills offer warranties against pile crushing or matting. These warranties, however, are subject to minimum pad thickness and pad density requirements. If these minimum requirements are not met, the no-mat portion of the warranty will become void. Therefore provide at least the minimum pad requirements for any carpet chosen.

Carpet installation

Now that all the proper materials have been selected, it is time to consider the most suitable installation method. There are two basic installation procedures: the stretch-in installation system and the direct glue-down method. Each procedure has unique benefits. The reasons for choosing one method over the other are predicated on the type of carpet to be installed and the conditions of the job site. For example, if the state of the subfloor is such that extensive repairs would be required to provide a suitable surface on which to glue down a carpet, then the more cost-effective approach may be simply to stretch in a carpet over a separate cushion, because stretch-in installations can often hide a multitude of subfloor irregularities. Another reason the stretch-in method may be recommended is because it increases both the R value and the noise reduction coefficients of any carpet installation. Also, the use of padding increases the life of a carpet because it inhibits crushing and matting of the surface pile.

Direct glue-down installations, on the other hand, are ideal in situations where, for example, carpet is to be installed on a ramp or where carts will be rolled across it. Large areas, in which a full stretch from one wall to another wall would be very difficult to achieve, are also perfect for glue-down installations. From a financial standpoint, the total price of a job may be decreased with a glue-down installation because the cost of the pad is eliminated. Buckling or rippling of the carpet is also minimized because there is no stretch to loosen after the carpet is laid.

Each method has advantages and disadvantages that must be considered in evaluating any given situation. However, regardless of the installation system selected, all job sites should be inspected to assess current conditions.

Preinstallation planning

It is always best to physically inspect each job site in order to make proper measurements and to check the existing subfloor and general conditions. If special care will be required to provide an appropriate surface on which to install the new carpet, the extent of that care should be noted and factored into the bid. Temperature and relative humidity on the site should be considered if there could be a problem as a result of these conditions. Fresh air ventilation requirements should be looked into and addressed, because during installation, and for at least 72 h thereafter, fresh air ventilation is necessary to eliminate any disagreeable odors.

If furniture or appliances are present, arrangements should be made between the owner and the installers to determine who is responsible for the removal and resetting of these items. If any doors need to be cut, which party is obligated to have them cut should be agreed upon. The disposal of any old carpet or the need for any new molding strips should be discussed as well. Anything that will influence the successful completion of the job should be contemplated ahead of time. The more issues that can be resolved prior to the installation date, the better.

Stretch-in carpet installation

The stretch-in method of installing carpet and pad is perhaps the most common technique used today. Almost all homes with carpeting have it installed in this fashion. The guidelines in this section are provided so you can get a general understanding of the procedures involved. Learning to install carpet takes years of dedicated apprenticeship and study. You must, however, familiarize yourself with these concepts in order to communicate intelligently with both your clients and installers.

Tackless strip

The proper stretch-in installation begins with the appropriate *tackless strip*. Tackless strip is a thin piece of plywood that is a minimum of 1 in wide and $\frac{1}{4}$ in thick. Into these plywood strips are inserted rows of sharp metal pins, positioned on an angle, onto which the carpet can be stretched and secured. These strips are also prenailed with anchoring nails. There are basically two kinds of tackless strip: one for wood floors and one for concrete floors. (If no anchoring nails are needed, a third kind of tackless strip has the preinstalled pins only. Since it cannot be face-nailed, this strip may be secured with construction glue or contact cement or by drilling and plugging nails into the substrate.)

(This latter type of tackless strip is highly successful when used over ceramic tile, metal floors, or terrazzo. If the subfloor is too hard, or if it is not advisable to nail into it for any reason, gluing the tackless strip is an option. For example, if a nail is driven into certain types of ceramic tile, the tile will merely shatter, and there will be no bond between the tile and the nail to hold the tackless strip in place. If the strip does hold temporarily, when the carpet is power-stretched onto it, the tackless strip will pull away from the tile, thus loosening the carpet.)

Tackless strip should be installed with the metal pins facing the wall. The distance between the wall and the tackless strip, called the *gully,* is typically a little less than the thickness of the carpet. However, in no event should it be over $\frac{3}{8}$ in from the wall. After the carpet is stretched onto the pins, the carpet is trimmed at the wall line and the remaining carpet is tucked into the gully.

Cushion installation

The *cushion* (carpet padding) chosen is cut into long lengths that are trimmed to within $\frac{1}{4}$ in of the tackless strip. The pad is then fixed to the floor with staples, if the floor is wood, or nonflammable adhesive, if the floor is concrete. The seams of the cushion are secured to each other with either staples, cushion adhesive, or tape (duct tape or specially designed pad tape). These procedures help keep the pad in place during installation; otherwise, the pad may wrinkle and create a visible lump underneath the carpet.

Carpet seaming

The ultimate success of any carpet installation depends on the appearance of the seams. If a diagram is provided for the layout of the job, refer to it for general seam placement. However, it is imperative that all measurements be rechecked *before* any carpet is cut. The old adage of *"Measure twice, cut once"* aptly applies to the installation of carpet. Since the cost of carpet is significant, even one wrong cut from a roll can adversely affect the outcome of a job.

The carpet is cut to slightly more than room size so it can be properly trimmed. All wrinkles are then removed. If the room is cold, the carpet should be allowed to acclimate for at least 24 h at a minimum temperature of 68° F before installation. Cold temperatures make the secondary back too stiff and installation extremely difficult. When carpet is stiff, it cannot be sufficiently stretched to achieve a taut condition.

Once the carpet is cut and in place, it may be necessary to seam two or more pieces together. There are two widely used methods for making seams: *row cut* and *straightedge*. To row-cut a seam, use an awl or a standard screwdriver to open up a row. Place the tool in a row about 1 in or less from the outer edge of the carpet, yet close to the selvage edge. Run the awl down that row the entire length of the carpet piece. Then follow in that furrow with a row-cutting tool (Fig. 3-19). A row cutter has a razor blade attached to it to make a clean, precise cut down the row. Use the same technique on the adjacent piece of carpet to achieve seam uniformity.

To straightedge a seam, turn the carpet face down and measure and mark the area on the carpet where the seam is to be. Use a chalk line to delineate the seam. Then with the aid of a 6-ft or 12-ft straightedge, cut on that line to make the seam. Make the same style cut on the adjacent piece, and then butt the two together.

Row cutting is a preferred method of seam cutting on Berber, commercial carpet, and patterned carpet. Straightedge seaming is more common on plush, tufted Saxony carpet, and the like. Check with the manufacturer of the carpet for the preferred seaming method if you have any doubts at all.

Due to modern-day high-speed manufacturing processes, the rows of the carpet may not be perfectly straight throughout the entire

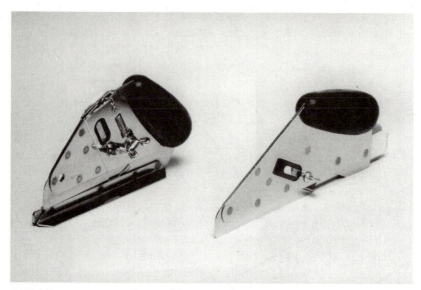

3-19 *Carpet row-cutting tools.* (Courtesy of Roberts Company.)

length of a long seam. Consequently, there may be areas along that seam where overlapping (or gaps) may appear. Sections that need to be closer will have to be kicked together with a knee-kicker and then *stay-nailed* (or *stay-tacked*) to temporarily keep them in place when seaming. Farther down the seam there may be an area where the carpet will need to be pushed out to properly join the two pieces together. These areas will also need to be stay-nailed to keep them temporarily secured. [Stay-nailing or stay-tacking is a technique of taking scrap strips of carpet, about 2 to 3 in wide, turning them upside-down a few inches or more away from the area you wish to secure, and then lightly, but not completely, hammering a few nails through both carpets into the subfloor (Fig. 3-20). This keeps the bottom carpet from moving during the seaming process.]

Please note that this stay-tacking technique can be used not just during seaming, but anytime you wish to temporarily secure a section of carpet. For example, a room currently has carpet installed in it, and you wish to change the doors from sliding glass doors to French doors. If you just pull up the carpet to change the doors, the entire room of carpet will lose its tautness and consequently will have to be completely restretched, because once a section of carpet is loosened, the whole carpet becomes relaxed. But, if you stay-tack a few feet into the room, allowing enough space to complete the door transition, the only area that will need to be restretched is the stay-tacked area. The rest of the carpet will maintain its tautness.

On most loop pile carpets, before two pieces are seamed together, a good-quality latex seam sealer should be applied to the edge of the carpet to prevent edge fiber loss, unraveling, fuzzing, or delamination at the seam. This seam sealer is smeared on the side edges, and then the two pieces are placed into each other as the seam is being made.

The most common methods of attaching two pieces of carpet that form a seam are hand sewing and hot-melt tape. Hand sewing is used today primarily for Axminsters, Wiltons, and most other woven carpets. But hand–sewing is bordering on becoming a lost art because many of the new installers use the hot-melt system almost exclusively. Most installers are not exposed to enough carpets that require hand–sewing, are inadequately trained to do so, or lack the patience it takes to make hand-sewn seams. Consequently, the number of qualified installers who can hand-sew a seam is dwindling.

The most frequently used seaming method, however, is the hot-melt tape system. This system consists of a fabric seaming tape (both 4- and 6-in-wide strips) that has a thermoplastic adhesive on its surface. This adhesive is heated by a specially designed electric iron that melts the adhesive and fuses the two pieces of carpet together (Fig. 3-21). The tape is placed directly underneath the seam. Beneath the tape is placed a firm board to help minimize any vertical movement. The iron is placed underneath the surface between the two

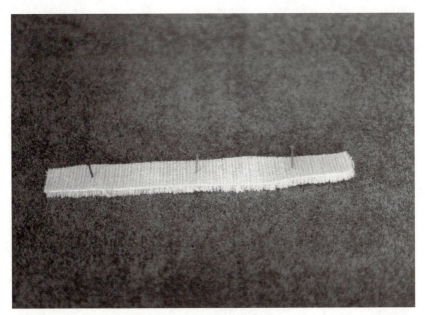

3-20 *Stay-tacked carpet.*

3-21 *Carpet seaming iron and seaming tape.*

pieces of carpet (directly on the seam tape) and is moved slowly down the entire length of the seam. As the iron melts the adhesive, the carpet edges are pressed together behind the iron as it makes its way down the row. This process permanently secures the two edges of carpet together (Fig. 3-22).

Carpet stretching

To obtain optimum tautness of a stretched-in carpet installation, it is essential to use a power stretcher (Fig. 3-23). A *power stretcher* is a device with a broad head and a metal plate. The head is composed of rows of angled metal pins. The pins are about $\frac{3}{4}$ to 1 in long. They can be adjusted up or down to accommodate varying carpet pile thickness. Attached to the head is a pole with a locking lever bar and a tail block. When pressed down, the locking mechanism pushes the carpet forward and locks it in place securely. The tail block is positioned against an opposite wall for support. This whole unit is approximately 3 ft long. Extension tubes can be added to increase the length of the power stretcher; 25 ft or more is not an uncommon length stretch. Once the carpet is stretched tight to the wall, it can be attached to the tackless strip. A knee-kicker and stair tool are then used to help secure the carpet to the pins of the tackless strip. A *stair tool* is a flat, wide steel instrument that resembles a scraper, but has a rounded edge so it will not cut into the carpet. It also has a thick, flat end that can take the force of a blow from a hammer, if necessary (Fig. 3-24).

A *knee-kicker* (Fig. 3-25) is a carpet-stretching apparatus. It is a smaller version of the power stretcher head, but with approximately

3-22 *Two pieces of carpet being seamed together.*

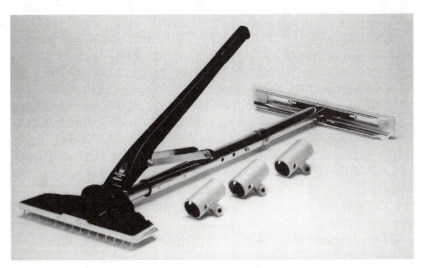

3-23 *Carpet power stretcher.* (Courtesy of Roberts Company.)

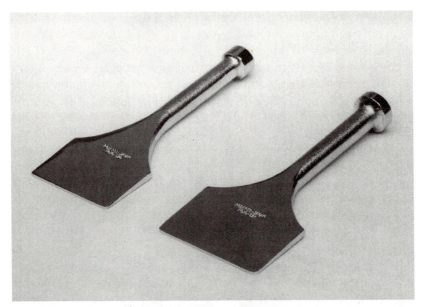

3-24 *Carpet stair tools.* (Courtesy of Roberts Company.)

four rows of long, adjustable metal pins as well as smaller pins in between the larger ones. It has a long tube that is often adjustable in
length. At the end of the tube is a square foam bumper pad. To use
the knee-kicker properly, carpet installers adjust the metal pins to the
appropriate pile height, insert the kicker into the carpet a few inches
from the wall, then kick the foam bumper with their knees to stretch
the carpet onto the tackless-strip pins.

The use of only a knee-kicker is not advisable for stretching an
entire room of carpet because an adequate stretch cannot be obtained without using a power stretcher. Surface traffic will tend to
loosen the carpet. The majority of a room should be power-stretched;
only certain portions should be kicked in. If a carpet is poorly
stretched, it may eventually develop ripples throughout the area.
These ripples, if they are walked on for any length of time, will destroy the integrity of the carpet by delaminating the secondary backing from the primary backing. So whenever ripples are noticed in a
carpet, have the entire room restretched immediately.

The manner in which a room is stretched in is very important.
The Carpet and Rug Institute (CRI), a national trade association rep-
resenting the carpet and rug industry, recommends that a room be
stretched in by following a series of specific steps (Fig. 3-26), designed to help achieve the best possible stretch. Figure 3-26 perfectly
illustrates those steps.

Note which steps require power stretching and which steps can be completed by using a only knee-kicker. The knee-kicker is used only in steps 3 and 5, after the carpet has been power-stretched. Also, note the angle of the stretches—all the main stretches are done at a *slight* angle, except that the last power stretch (step 8) is *straight*.

Once the carpet has been stretched in place, the excess carpet at the walls can be trimmed with a specially designed tool called a *carpet wall trimmer* (Fig. 3-27). A wall trimmer neatly trims the carpet the correct distance from the wall, leaving enough carpet to be tucked into the gully.

To complete the installation, trim the seam with carpet napping shears (Fig. 3-28). These scissors have a raised, offset handle that keeps one's hand elevated from the carpet. As a result, napping shears allow for easy, level cutting. This procedure will clip off any raised strands of yarn and give the seam a well-groomed appearance.

With everything in place, all that is left to do is to vacuum the carpet and clean up any debris created during the installation. It is always best to leave the job site in pristine condition. All customers appreciate a finished job that is presented to them in an immaculate state.

Direct glue-down carpet installation

Carpet can be glued directly to an existing substrate when a stretch-in installation is either not desired or not practicable. The secondary backings that can be glued down are polypropylene, woven, unitary, and foam backings. Although used less often in residential settings,

3-25 *Carpet knee-kicker.* (Courtesy of Roberts Company.)

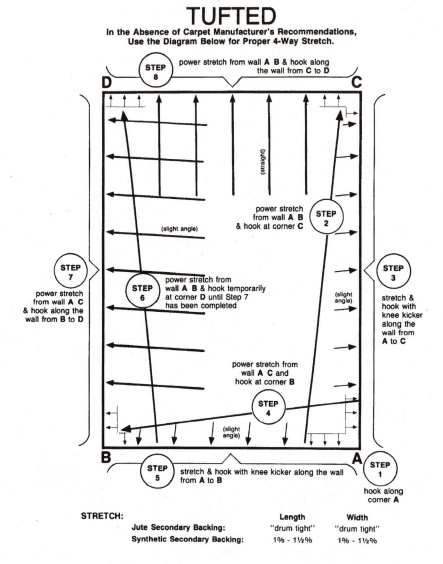

3-26 *Proper steps for stretching a room of wall-to-wall carpeting.*
(Courtesy of Carpet and Rug Institute.)

glue-down installations are more widely used in commercial situations. The preparation of the existing floor surface is more critical on a glue-down job than on a stretched-in one, so careful inspection of the subfloor is necessary.

    The substrate must be in a condition suitable to accept the new adhesive in order for a secure bond to be created between the carpet and the subfloor. The subfloor must be clean, dry, and smooth. It should also be free of paint, waxes, and oils. Any dirt or dust has to be sufficiently swept away; otherwise, the adhesive will not perform adequately because the presence of dirt and dust can cause adhesive bond failure.

3-27 *Carpet wall trimmers.* (Courtesy of Roberts Company.)

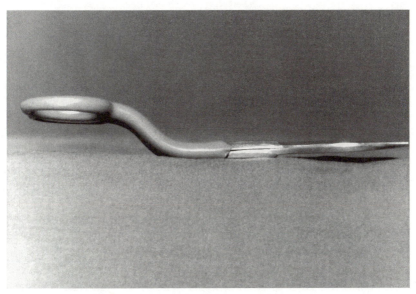

3-28 *Carpet napping shears.* (Photo by Mahendra Ramawtar.)

Any cracks in the surface should be filled with an appropriate patching compound to provide a flat, smooth surface. If any old floor coverings need to be removed, be sure to properly clean off any residual adhesive from the subfloor. Older adhesives are sometimes incompatible with the new adhesives and may create adhesion problems.

Once the floor has been adequately prepared, it is important to apply the appropriate adhesive with the correctly notched trowel recommended by the carpet manufacturer. Keep in mind that the notches on a trowel begin to wear down with use and may therefore need to be renotched (or changed entirely) to maintain the appropriate depth of adhesive spread. Different types of backing materials require different trowel notching. For example, carpets that have coarse backings require larger notched trowels. So by paying particular attention to the proper trowel notching, complete adhesive transfer into the carpet backing will be achieved. If an insufficient amount of adhesive is applied to the substrate, the carpet will wrinkle and lose its bond.

After the seams have been trimmed and are ready to be installed, spread the adhesive; then, according to the adhesive manufacturer's guidelines, allow the proper *open time* (the time necessary for the adhesive to set up and be ready to accept the new carpet). Now, if more than one breadth of carpet is to be installed, firmly set the first piece in place. To prevent the carpet from unraveling or fraying, apply a bead of latex adhesive along the edge of the first piece of carpet. When ready, slide the second carpet edge into the first edge. This will bond the two edges together.

After the carpet is installed, roll the carpet with a linoleum roller (Fig. 3-29) to obtain the maximum adhesive transfer. A 60- to 100-lb roller should be sufficient. Then, to complete the job, trim the carpet at the walls and use the napping shears to groom the seam. After that, vacuum the carpet and pick up all the loose scraps. The job is completed.

Modular carpet tiles

Another type of carpet product that can be glued down is *modular carpet tiles*. Modular carpet tiles are typically made by fusing either loop pile or cut-pile yarns onto a polyvinyl chloride (PVC) backing material. These carpet tiles generally come in squares that are 18 in × 18 in.

Modular carpet tiles are used primarily for commercial, rather than residential, purposes. These products possess such an incredible tuft bind that edge ravel of the yarn is not a problem. There are two main advantages to installing carpet tiles, rather than broadloom carpet: In large office spaces, an installation can be accomplished by doing small sections at a time, and repairs to a damaged tile can be done

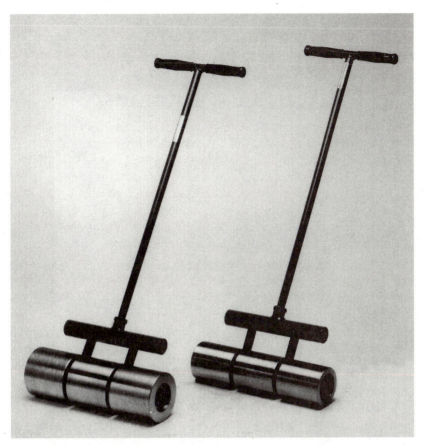

3-29 *Linoleum rollers.* (Courtesy of Roberts Company.)

rather easily. For example, if a heavily used walkway begins to show excessive wear, those tiles can be replaced in a matter of moments without disrupting the entire office.

The type of adhesive often specified for installing carpet tiles is what's known as a *release adhesive*. A release adhesive snugly secures the tile to the substrate, yet when the need arises, the tile can be easily lifted from the surface of the floor. This makes tile replacement a snap.

Modular carpet tiles come in a variety of patterns and textures. There are cut-pile solids and cut-pile graphics, as well as level loop heathers and level loop graphics. Carpet tiles can also provide unique designs by mixing and matching various colors. Borders and artful motifs can be taken to the limits of the imagination.

Carpet tiles are somewhat of a premium-priced product, so they are not necessarily suitable for all commercial installations. However, whenever a situation presents unusual challenges, from either an installation or a design standpoint, remember the incomparable way in which modular carpet tiles can solve these problems. They provide you with yet another tool in your repository of flooring resources.

Carpet warranties

Since a great majority of the carpet produced today has some type of written performance assurance, it is important to be aware of the warranties that accompany these products. For the purposes of this discussion, we will concentrate on those warranties regarding nylon fibers. Because carpet is more susceptible to staining and soiling than other floor covering materials, such as tile or resilient flooring, carpet fiber manufacturers have developed yarn systems that are more likely to better resist these problems. It is important to know that nearly all the written warranties specifically state that no carpet is fully "stain-proof." Manufacturers realize that carpet may stain, yet they maintain that the carpet should clean up readily because of the advanced technological breakthroughs in fiber treatment procedures.

In addition to the stain and soil warranties, there are warranties that cover surface pile abrasive wear, antistatic warranties that guard against static shock, and texture retention warranties that protect against yarn fiber untwisting. (When a carpet fiber untwists, it causes the yarn to mat down. A texture retention warranty is also referred to as a *no-mat, no-crush warranty.*)

Keep in mind that all manufacturers have their own carefully worded warranties, and the language for a specific carpet warranty should be read thoroughly to learn what is covered and, more precisely, to understand what is *not* covered. Sometimes knowing what an agreement does not cover can be more helpful than being familiar with all the items that are covered. Consult the manufacturer's warranty brochure for the exact wording of each and every product you offer.

In general, almost all nylon warranties will exclude coverage for damage caused by acne medicines. These products contain benzoyl peroxide which can destroy most dyestuffs (color) in the carpet. An oxidizing agent, acne medicines cause the affected areas of the fiber to become white with pink, yellow, or orange highlights. Therefore, if consumers know about this exclusion, those in the home using acne medications should be made aware that extreme care need be taken so that no medicine comes in contact with the carpet. One of the more common ways for these medicines to be transferred to a carpet occurs when people apply them to their skin and then fail to wash their hands to remove the residual ointment. If they were to then lay down on the carpet to watch television, there is a good possibility the carpet could become contaminated if they rub their hands on it.

In addition to the warranties provided by the fiber producers, the carpet manufacturer—the company that actually tufts and finishes the carpet—may include additional warranties. If the fiber is not a *branded fiber* from a fiber producer, the carpet manufacturer may include its own stain, soil, or wear warranties. In those instances, all claims will be handled strictly by the carpet manufacturer. It is very important to know the distinction between a fiber producer's warranty and a carpet manufacturer's warranty. Regardless of who warrants the performance characteristics of the carpet, many of the items covered, and the limitations thereof, are similar.

Soil and stain resistance

Almost all the warranties against staining on nylon fibers typically refer to stains caused by common "household food and beverage" products. Fibers that have advanced stain-resistant qualities are designed to resist these stains better than equivalent untreated carpet.

Stain resistance is the capacity of a fiber to resist permanent staining. Since the fiber producers have extensively tested the performance of these materials, they know which substances or occurrences will adversely affect the yarn. Substances that are known to create unfavorable results are specifically excluded from coverage. Some of these substances include, but are not limited to, the following: plant foods, acne medication, bleaches, insecticides, drain cleaners, mustard, herbal teas, vomit, urine, feces, iodine, and similar substances. Improper maintenance or poor cleanup procedures can also void the warranty.

All the fiber producers provide pamphlets outlining cleaning instructions that clearly explain how to treat specific stains. Both the cleaning agents and the sequence of their use are identified (Fig. 3-30). If you follow these instructions, the warranty should remain intact.

The soiling of a carpet fiber differs from staining in that soiling is an accumulation of dirt over time, while staining is done by accident. Soiling is dirt that attaches itself to the carpet fiber, and as a result, the carpet begins to have a lackluster appearance. Most manufacturers of nylon carpet recommend that the carpet be cleaned using a *hot water extraction* (steam-cleaning) method. Other methods such as a dry foam or powder or rotary shampooing can also be used, but hot water extraction seems to be the preferred choice.

Regular vacuum cleaning is an important factor in keeping a carpet looking fresh and attractive. Vacuuming is also critical in removing

Spot Removal Chart

SOLUTIONS
SOLUTIONS
SOLUTIONS
SOLUTIONS
SOLUTIONS
SOLUTIONS

	Nail Polish Remover	Dry Cleaning Fluid	Detergent Solution	Warm Water Rinse	White Vinegar Solution	Ammonia Solution	Spot Removal Kit	Call a Professional
Alcoholic Beverages			1	4	3	2		5
Blood*			2	3		1	4	5
Candle wax	1	2						3
Chewing gum*		1						2
Chocolate		1	2	5	4	3		6
Coffee			1	3	2		4	5
Crayon		1	2					3
Dye	3	1	2					4
Fingernail Polish	1		2	3				4
Band lotion		1	2	4		3		5
Ice cream		1	2	5	4	3		6
Ink (ballpoint)*	1	2	3	6	5	4	7	8
Kool-Aid®			1	3	2		4	5
Latex paint			1	3		2	4	5
Lipstick	1	2	3	6	5	4	7	8
Mustard			1	3	2		4	5
Rust*			2	3	1		4	5
Shoe polish, paste	1	2	3	4				5
Soft drinks			1	4	3	2	5	6
Tomato sauces			1	3	2		4	5
Unknown	2	1	3	6	4	5	7	8
Urine			1	2	3	4		5
Vomit			1	4	3	2	5	6

Perform steps in order (1,2,3, etc.) while following the Spot Removal Directions.

*See additional spot removal suggestions (Effective Treatment For Difficult Spots).

3-30 *Carpet cleaning instructions.* (Courtesy of Carpet and Rug Institute.)

SPOT REMOVAL SOLUTIONS

Vacuum - Vacuum all dry spills to lift and remove as much of the substance as possible. Pouring any liquid into a dry spill may create additional problems.

Dry Cleaning Fluid - A non-flammable spot removal solution is preferred. Exercise caution when using a dry cleaning fluid. Never pour a dry cleaning solution directly onto the carpet or allow it to reach the backing. Dry cleaning fluids may destroy the latex adhesive that holds primary and secondary backings together. Apply Dry cleaning fluids to a cloth and blot.

Nail Polish Remover - Two types of nail polish removers are available. One type contains acetone, a dry cleaning solvent. When using nail polish removers containing acetone, use the same precautions as with other dry cleaning solutions. The second type contains amyl acetate or ethyl acetate, which is used in many paint, oil and grease (POG) removers. Many POG removers leave residues that may cause rapid soiling. When using a POG remover always rinse the area thoroughly with a dry solvent to remove residues. (See residue precautions)

Detergent Solution - Mix ¼ teaspoon of a mild, liquid dishwashing detergent per cup of lukewarm water. NEVER USE A STRONGER CONCENTRATION! Thorough rinsing is necessary to remove detergent residues that may cause rapid soiling. It may be necessary to rinse with warm water several times to completely remove residues (See Residue Precautions). Use care in selecting a detergent. Do not use an automatic dishwashing liquid! Many contain bleaching agents that destroy carpet color. Never use a laundry detergent of any type, because they may contain optical brighteners (fluorescent dyes) that dye carpet fiber.

Warm Water Rinse - Use lukewarm tap water in most cases to rinse the cleaning solutions from fibers. Failure to completely rinse solutions from the fiber may cause accelerated soiling. A spray bottle works particularly well in rinsing the area without overwetting.

Vinegar Solution - Mix 1 cup of white vinegar per 2 cups of water. White vinegar is a 5% acetic acid solution, and is used most often to lower the alkalinity caused by detergent solutions or alkaline spills (see Spot Removal Chart).

Ammonia Solution - Mix 2 tablespoons of household ammonia per cup of water.

Spot Removal Kit - Available from most carpet retail stores. Follow directions closely! Some spot removal kits contain: (A) a detergent solution and (B) a stain resist solution. Use of the (B) stain-resist solution prior to the complete removal of the spill may cause a permanent stain. Other spot removal kits may contain a dry extraction cleaning compound that can be used effectively for most common household stains.

Call a Professional - Carpet cleaning professionals have the ability and the equipment to use more aggressive cleaning solutions to remove stubborn spills. Always consider consulting a professional regarding any spot removal question.

3-30 *Continued.*

surface dirt before it can penetrate deeply into the carpet pile. This lack of pile penetration by foreign substances, from both a soiling and staining standpoint, is what makes the modern nylon fibers so superior. These adverse substances remain higher up on the pile, for a longer time, which gives the owner a better opportunity to clean up the stain or soil. The more deeply, and quickly, a spill or dirt is absorbed into the fiber, the more difficult it is to remove.

Most manufacturers will not even entertain a claim for soiling or staining until after it has been cleaned at least once by a qualified professional carpet cleaner. Some companies go even further by stating that the carpet should be cleaned at least once a year. Therefore, it is important to follow the manufacturers' directions and guidelines carefully. All general conditions and stipulations should be taken into consideration before you attempt to treat any questionable situation.

Every fiber manufacturer and carpet producer has an 800-number phone service available for consumers to call regarding stain problems or other warranty issues. They will gladly assist consumers by providing information pertaining to the cleaning of any specific spill or stain.

Wear warranties

Carpet wear warranties are offered by either the carpet or fiber manufacturer on most carpets produced today. *Wear* is the amount of pile fiber loss—or yarn weight loss—caused by abrasive wear. Normal abrasion of a carpet fiber is a result of surface traffic.

These warranties are for a specified number of years, usually 5, 10, or 20. Also, the pile fiber loss is not to exceed a certain percentage of the original fiber content. Therefore, a typical wear warranty will state, e.g., that the carpet fiber will not abrasively wear away more than 10 percent in 10 years. This means that in 10 years, 90 percent of the fiber content should remain intact; if not, the consumer can file a claim with the company warranting the performance of the carpet. (Carpet on stairs, however, is always excluded from any wear warranty.)

It is necessary to point out that wear and appearance are two entirely different concepts when it comes to this section of the warranty. *Wear* means fiber loss, whereas *appearance* means the look of a carpet. The appearance aspect of a carpet warranty falls under the texture retention portion of the agreement.

Texture retention warranties

Since the texture retention warranties refer to the ability of the yarn to resist crushing and matting, these are significant sections of the warranty to consider. This portion tells you how the carpet will perform and look, in the years covered. Most texture retention warranties are for 5 or 10 years, yet some are for longer. *Crushing* refers to loss of pile thickness caused by compression, while *matting* refers to the untwisting and intermingling of the yarn. This untwisting causes the fiber shaft to lose its ability to spring back and so causes the yarn to simply lie flat. Since most carpets begin to lose their original appearance more quickly than they wear out, this feature is very significant.

Antistatic warranties

Annoying static shock from carpet fibers has been virtually eliminated in these new, advanced fiber systems. Most products include limited or lifetime antistatic provisions.

Warranties on other fibers

Olefin (polypropylene), acrylics, and polyester fibers carry their own particular warranties. Since they have a different chemical makeup from that of nylon, certain substances may affect them in decidedly different manners. For example, substances that permanently stain a nylon fiber may be readily cleaned from an olefin fiber and vice versa. Therefore, consult the warranty brochure on any product to find out its distinguishing characteristics.

Carpet care

To keep a carpet in optimum condition, a regular maintenance program is essential. Vacuum cleaning of the main thoroughfares in a home or an office should be done on a daily basis; a minimum of at least twice a week is recommended for the entire area. Professional cleaning should remove almost all embedded soil. Spot cleaning with the recommended products removes most stains quickly. Topical treatments, such as stain repellents or soil retardants, should be applied only after you consult the carpet manufacturer. Aftermarket application of these products could void any original warranties if the treatment is not an approved procedure or brand.

4

Wood flooring

From prehistoric times to the present, people have used wood products to provide many of the basic necessities of life. Wood has been an excellent source of fuel as well as a superb material for constructing houses to provide shelter. It would be difficult to imagine the world without trees or the products derived from them.

As a floor covering material, wood has been used since biblical days and beyond. Early recorded history makes numerous references to wood flooring. There are many wood floors in palaces and castles throughout Europe that are several centuries old. To see these floors today, one must marvel at their fine condition. The beauty and functional characteristics of wood are immeasurable. The unique texture and grain pattern of a wood floor are almost like a work of art. Each individual piece of wood is like a human fingerprint—singular, rare, and distinct. No two pieces of wood are identical, and therein lies its most beautiful aesthetic feature.

Since wood is a natural by-product of trees, it has inherent properties that require special attention. Once a growing structure composed of living cells, when felled, trees (or wood) still react acutely to the influences of atmospheric humidity. Understanding the essential traits of wood, from its existence as a living, dynamic entity until it is laid in place as a floor covering, will make comprehending that transition simple.

Composition of wood

Wood is very similar in composition to a sponge that has been taken from the sea. It has a distinct cellular structure that is extremely porous, and the amount of water it can retain is enormous. Unlike the sponge, however, in a tree it is difficult to extract water by direct compression. Only elapsed time or special drying methods can generally remove water from wood. Wood's spongelike ability to attract and evaporate moisture makes it a somewhat unpredictable material to work with because of its changeable nature. Yet, thinking of wood as a "sponge" will help you to understand its behavioral patterns. Although there are a great many varieties of trees, they all share many common features. Trees are perennial plants that have the capability of adding new growth to prior growth every year. The trunk of the tree provides the majority of the wood used for flooring. A cross section of a trunk shows its typical characteristics (Fig. 4-1).

Figure 4-1 is an illustration of an oak trunk. The outermost portion of the trunk is the bark. The bark consists of two separate layers: the *rhytidome* (outer bark) and the *phloem* (inner bark). The rhytidome comprises tissue that is dead, while the phloem is the living section of the bark. It is the phloem that provides nutrition to the tree through *photosynthesis.*

Photosynthesis is a process in all plants (including trees) whereby carbon dioxide from the atmosphere and water from the leaves of the plant are combined to form carbohydrates in the chlorophyll-containing tissues that have been exposed to sunlight. This series of events produces sap, which is a basic sugar that the tree uses as nourishment for its own development. The sap carries nourishment to the

INNER BARK
(Phloem) carries sugar made in the leaves down to the branches, trunk, and roots, where the sugar is made into other substances vital for growth.

SAPWOOD
(Xylem) carries sap (water and dissolved minerals and nitrogen) from roots to leaves, where sugar is made. The summerwood is usually denser than the springwood.

OUTER BARK
Protects tree from injuries by animals, harmful insects, diseases, excessive heat, cold, dryness etc.

HEARTWOOD
Gives extra strength and stiffness (was sapwood, but now is inactive.)

CAMBIUM
A single layer of living cells between bark and wood where growth in diameter occurs. It forms annual rings of wood inside and new bark outside.

4-1 *Cross section of tree trunk showing its various elements.* (Illustration courtesy of the U.S. Forest Service.)

cambium layer (a layer of living cells) where new cells are created. The miracle of photosynthesis produces two wonderful events that are truly remarkable: It not only adds oxygen to the air, but also removes carbon dioxide from the atmosphere. That is why we often hear that trees are the lungs of the earth.

The cambium is the area of the tree where growth in diameter actually takes place. The cambium layer separates the bark from the wood. The function of these living cells is to divide and create more cells for development, because the existing cells do not by themselves cause growth. It is the cambium that actually forms the annual growth rings.

The section next to the cambium is the actual wood itself. The wood consists of two zones: sapwood and heartwood. Sapwood (xylem) is the area of the tree that contains some of the living elements of the organism. These cells carry fluids from the roots, through the trunk, to the leaves and upper segments of the tree. Sapwood and the cambium hold vast amounts of fluids, often far surpassing their own dry weight. It is not uncommon for a tree to be swollen inside with water. This is the most important concept to grasp when you study tree growth. The water content of a tree, and subsequently of its resulting lumber, will be of great concern when you install a wood floor.

When sapwood is no longer useful as a sap-conducting vehicle, it becomes heartwood. The cells within the heartwood are not active cells. Heartwood is often darker than sapwood due to deposits of chemicals within the tree's structure that are toxic to fungi and insects.

The *pith* marks the point that is the very center (the beginning, if you will) of the tree. Annual *growth rings* circle the pith in bands that are alternately light and dark. These colored bands are known as *earlywood* and *latewood,* depending upon the time the actual growth occurs in the growing season. Earlywood and latewood are also different in density, which accounts for how each band accepts a staining product, once it is installed as a flooring material.

Wood classifications

There are two categories of wood: hardwoods and softwoods. Hardwoods come from trees that are *deciduous* (broadleaf trees that lose their leaves every winter). Softwoods are from trees that are commonly referred to as *conifers* (needle-bearing trees).

One of the main differences between the two is that hardwoods are porous and softwoods are nonporous. Hardwoods have pores (or vessels) whereas softwoods do not. Pores in wood are actually holes in the cells. Wood cells are elongated and tend to be oriented in one direction. Examples of hardwoods are oak, maple, mahogany, and walnut. Certain softwoods are fir, cedar, spruce, and pine.

The most significant indicator of strength within these different types of wood is the density. A wood's density determines its hardness and resistance to nailing. A wood that has a greater density, such as oak, will make a much better floor covering product than a wood with a lesser density, such as cedar. Woods that are less dense tend to dent more easily because they are soft. The cell structure of a wood creates its density. A strong wood has a dense cell structure, while a softer wood has a less compressed cell structure. A floor that can withstand the rigors of everyday wear and tear will be one with a dense cell structure.

Moisture content

The living elements of a tree are bloated with excessive water in any nondrought environment. Once the tree is cut to produce boards for flooring, the water begins to evaporate. This is the point at which wood begins to shrink. The moisture content of wood is influenced by the humidity in the air. For wood to be a product with less changeability, there must be a balance between the moisture content of the wood and the humidity in the atmosphere. The dimensional size of the precut boards can change directly as a result of the humidity in an environment. Therefore, wood purchased for a flooring job should be specified as *preshrunk,* to allow it to acclimate to the job site's atmospheric conditions.

Wood is *hygroscopic,* which means that it readily takes up and retains moisture. Since wood is so greatly influenced by the relative humidity (RH) in the air, the *bound water* (water that is retained in the cell walls) reacts according to the changes in pressure. Ultimately, at some point a piece of wood stops having fluctuations in moisture content at a given RH level. This is called the *equilibrium moisture content* (EMC) (Fig. 4-2). Study Fig. 4-2 to see how relative humidity affects the moisture content.

The use of a moisture meter is good insurance against installing a wood floor with an excessive moisture content. Wood flooring should have a moisture content that is between 6 and 9 percent. When readings are higher than that, the floor materials should be allowed to dry further before installation is begun.

Solid wood flooring should be brought to the job several weeks before installation. The wood should be left uncovered so it can ac-

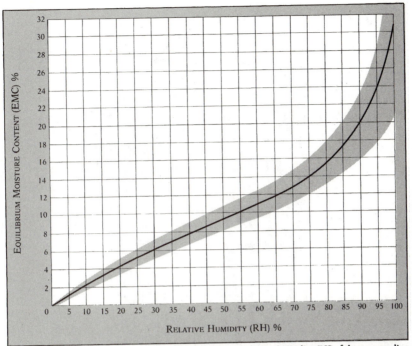

1—The amount of bound water in wood is determined by the relative humidity (RH) of the surrounding atmosphere; the amount of bound water changes (albeit slowly) as the relative humidity changes. The moisture content of wood, when a balance is established at a given relative humidity, is its equilibrium moisture content (EMC). The solid line represents the curve for white spruce, a typical species with fiber saturation point (FSP) around 30% EMC. For species with a high extractive content, such as mahogany, FSP is around 24%, and for those with low extractive content, such as birch, FSP may be as high as 35%. Although a precise curve cannot be drawn for each species, most will fall within the color band.

4-2 *Equilibrium moisture content.* (From Understanding Wood by R. Bruce Hoadley used with permission of the Taunton Press, Inc. 63 South Main Street, PO Box 5506, Newtown, CT 06470. © 1980 The Taunton Press, Inc. All rights reserved.)

climate to the specific conditions of the location. In addition, the subfloor moisture content should be checked. The variance between the subfloor EMC and that of the wood flooring should not exceed 4 percent. If it does, corrective measures must be taken either to the subfloor or to the wood itself to bring the percentages closer to an acceptable level.

When you install the floor, it is necessary to leave a gap around the perimeter of the room between the flooring and the wall to allow for any expansion and contraction that may occur later as a result of the wood's gaining or losing moisture. This gap is called an *expansion gap*. After the floor is laid, the expansion gap is covered up by a baseboard or shoe molding. Any subsequent movement in the floor will occur underneath these wall moldings and therefore will go unnoticed by the inhabitants.

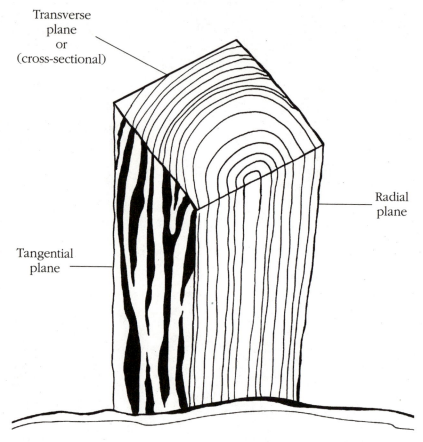

Transverse
plane
or
(cross-sectional)

Radial
plane

Tangential
plane

4-3 *The three planes in wood.*

Planes in wood

Wood can be cut on one of three planes: transverse (cross-sectional), radial, or tangential (Fig. 4-3). The appearance of the wood on the three planes is a result of growth patterns (rings) and the elongated cell orientation of the wood itself.

The *transverse* plane is the plane that would be visible if the tree were simply sawed off at its base. The remaining circular tree stump top would reveal the transverse plane. The *radial* plane runs perpendicular to the growth rings and produces wood with a very tight grain pattern because of the elongated growth lines. The *tangential* plane, in turn, is perpendicular to the radial plane. This plane is tangent to the circle created by the stump. The tangential plane yields wood with the most pleasing grain pattern because the plane cuts along the irregular curvature of the growth rings. Each of these planes will produce a board that has a different grain pattern. Therefore, how a log is sawed will affect the look and grain of the wood.

Sawing patterns in wood

A log can be cut in a number of patterns. Each pattern will produce varying grain configurations. The most common patterns are *rift-sawed, quarter-sawed,* and *flat-sawed* (Fig. 4-4). Rift sawing highlights a log's vertical grain since the broad face is more in the radial plane. Quarter sawing is a variation of rift sawing since quarter-sawed pieces have a surface that is also mainly radial on the face of the board. Flat-sawed (Fig. 4-5) (or *plain-sawed*) boards have a face that is primarily on the tangential plane. This form of sawing pattern is the most common because it produces the most diverse grain patterns.

Once a tree is sawed into boards, the boards must be evaluated and graded. Since some boards will still have bark (wane) on them, they must be cut to eliminate any unsightly features. A look at the

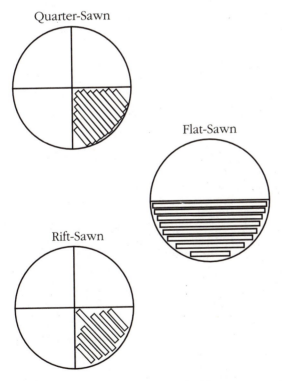

Quarter-Sawn

Flat-Sawn

Rift-Sawn

4-4 *Three patterns of sawing lumber.*

4-5 *A flat-sawed board.*

lumber evaluation chart (Fig. 4-6) will help you visualize how a board is properly trimmed.

Hardwood flooring grades

Solid-hardwood flooring (as opposed to softwoods or laminated wood flooring products) have a grading system published by *National Oak Flooring Manufacturers Association* (NOFMA). The grading rules for unfinished flooring classify boards into groups with similar characteristics: *clear, select, number one common,* and *number two common* (Fig. 4-7). The chart classifies unfinished as well as prefinished products. Unfinished oak flooring is further broken down into plain-sawed (flat), rift-sawed, and quarter-sawed products. Other woods included in this classification chart are hard maple, beech, birch, and pecan.

Study the chart and pictures carefully to become aware of the differences in grades. Note that clear and select have excellent appearance qualities. They can also be combined to produce a high-quality, cost-effective wood floor. Number one common has a more variegated appearance with knots and holes that will require filling. Number one common is still a fine floor that will have a great deal of grain

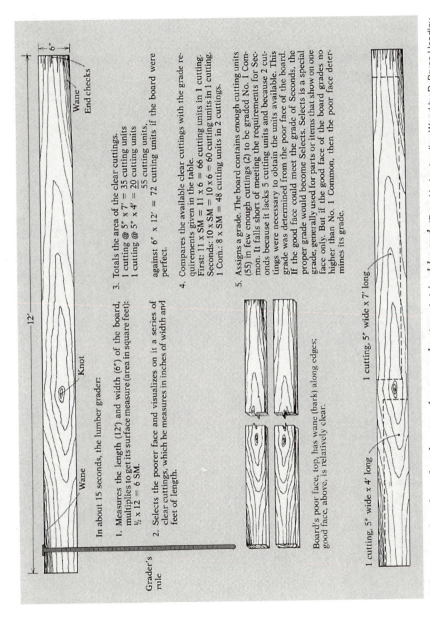

In about 15 seconds, the lumber grader:

1. Measures the length (12') and width (6") of the board, multiplies to get its surface measure (area in square feet): $\frac{1}{2} \times 12 = 6$ SM.

2. Selects the poorer face and visualizes on it a series of clear cuttings, which he measures in inches of width and feet of length.

Board's poor face, top, has wane (bark) along edges; good face, above, is relatively clear.

3. Totals the area of the clear cuttings.
1 cutting @ 5' x 7' = 35 cutting units
1 cutting @ 5' x 4' = 20 cutting units
                      55 cutting units,

against 6" x 12' = 72 cutting units if the board were perfect.

4. Compares the available clear cuttings with the grade requirements given in the table.
First: 11 x SM = 11 x 6 = 66 cutting units in 1 cutting.
Seconds: 10 x SM = 10 x 6 = 60 cutting units in 1 cutting.
1 Com.: 8 x SM = 48 cutting units in 2 cuttings.

5. Assigns a grade. The board contains enough cutting units (55) in few enough cuttings (2) to be graded No. 1 Common. It falls short of meeting the requirements for Seconds because it lacks 5 cutting units and because 2 cuttings were necessary to obtain the units available. This grade was determined from the poor face of the board. If the good face could meet the grade of Seconds, the proper grade would become Selects. Selects is a special grade, generally used for parts or items that show on one face only. But if the good face of the board grades no higher than No. 1 Common, then the poor face determines its grade.

1 cutting, 5" wide x 4' long

1 cutting, 5" wide x 7' long

4-6 *Steps taken by a lumber grader to evaluate a typical board.* (From Understanding Wood by R. Bruce Hoadley used with permission of the Taunton Press, Inc. 63 South Main Street, PO Box 5506, Newtown, CT 06470. © 1980 The Taunton Press, Inc. All rights reserved.)

121

UNFINISHED OAK FLOORING (Red & White Separated)	UNFINISHED HARD MAPLE (BEECH & BIRCH*)	UNFINISHED PECAN FLOORING*	PREFINISHED OAK FLOORING (Red & White separated, graded after finishing)
CLEAR PLAIN or CLEAR QUARTERED* Best appearance. Best grade, most uniform color, limited small character marks. Bundles 1¼ ft. and up. Average length 3¾ ft. **	**FIRST GRADE** Best appearance. Natural color variation, limited character marks, unlimited sap. Bundles 1¼ ft. and up. 1¼ ft., 2 ft., & 3 ft. bundles up to 45% footage. 2 ft. bundles up to 25% footage. 1¼ ft. bundles up to 5% footage.	**FIRST GRADE** Excellent appearance. Natural color variation, limited character marks, unlimited sap. Bundles 2 ft. & up. 2 & 3 ft. bundles up to 25% footage. **FIRST GRADE RED** (Special Order) **FIRST GRADE WHITE** (Special Order)	**PRIME GRADE (Special Order)** Excellent appearance. Natural color variation, limited character marks, unlimited sap. Bundles 1¼ ft. & up. Average length 3½ ft.
SELECT PLAIN or SELECT QUARTERED* Excellent appearance. Limited character marks, unlimited sound sap. Bundles 1¼ ft. and up. Average length 3¼ ft. ** **SELECT & BETTER*** A combination of Clear and Select grades. SELECT AND BETTER	**SECOND GRADE** Variegated appearance. Varying sound wood characteristics of species. Bundles 1¼ ft. and up. 1¼ ft., 2 ft., & 3 ft. bundles up to 55% footage. 2 ft. bundles up to 27% footage. 1¼ ft. bundles up to 10% footage. **SECOND & BETTER GRADE** A combination of First & Second Grades. Lengths equivalent to Second Grade. SECOND AND BETTER	**SECOND GRADE** Variegated appearance. Varying sound wood characteristics of species. Bundles 1¼ ft. and up. 1¼ ft. to 3 ft. bundles as produced up to 40% footage. **SECOND & BETTER GRADE** A combination of FIRST and SECOND GRADES.	**STANDARD GRADE** Variegated appearance. Varying sound wood characteristics of species. A sound floor. Bundles 1¼ ft. & up. Average length 2¾ ft. **STANDARD & BETTER GRADE** Combination of STANDARD and PRIME. Bundles 1¼ ft. & up. Average length 3 ft. STANDARD AND BETTER
NO. 1 COMMON Variegated appearance. Light and dark colors; knots, flags, worm holes and other character marks allowed to provide a variegated appearance after imperfections are filled and finished. Bundles 1¼ ft. and up. Average length 2¾ ft. **	**THIRD GRADE** Rustic appearance. All wood characteristics of species. Serviceable economical floor after filling. Bundles 1¼ ft. and up. 1¼ ft. to 3 ft. bundles as produced up to 75% footage. 1¼ ft. bundles up to 45% footage.	**THIRD GRADE** Rustic appearance. All wood characteristics of species. A serviceable, economical floor after filling. Bundles 1¼ ft. and up. 1¼ ft. to 3 ft. bundles as produced up to 60% footage. **THIRD & BETTER GRADE** A combination of FIRST, SECOND and THIRD GRADES.	**TAVERN GRADE** Rustic appearance. All wood characteristics of species. A serviceable, economical floor. Bundles 1¼ ft. & up. Average length 2¼ ft. **TAVERN & BETTER GRADE** (Special Order) Combination of PRIME, STANDARD and TAVERN. All wood characteristics of species. Bundles 1¼ ft. & up. Average length 3 ft. TAVERN AND BETTER
NO. 2 COMMON Rustic appearance. All wood characteristics of species. A serviceable economical floor after knot holes, worm holes, checks and other imperfections are filled and finished. Bundles 1¼ ft. and up. Average length 2¼ ft. ** Red and White may be mixed.	**THIRD & BETTER GRADE** A combination of First, Second & Third Grades. Bundles 1¼ ft. and up. 1¼ ft. to 3 ft. bundles as produced up to 60% of footage. THIRD AND BETTER		
1¼' SHORTS Pieces 9 to 18 inches. Bundles average nominal 1¼ ft. **NO. 1 COMMON & BETTER SHORTS** A combination grade of CLEAR, SELECT, & NO. 1 COMMON **NO. 2 COMMON SHORTS** Same as No. 2 Common.			**PREFINISHED BEECH & PECAN FLOORING** **TAVERN & BETTER GRADE** (Special Order) Combination of PRIME, STANDARD and TAVERN. All wood characteristics of species. Bundles 1¼ ft. & up. Average length 3 ft.

(NOTE: Flooring specified "QUARTERED" shall contain Quartered only. Flooring specified "PLAIN" may contain both Plain and Quartered.)
*Check for availability.

**The average lengths for 7' and shorter packages, nested or otherwise, are as follows:

CLEAR	3½'	#1COMMON	2¾'
SELECT	3'	#2COMMON	2'

A brief grade description, for comparison only NOFMA flooring is bundled by averaging the lengths. A bundle may include pieces from 6 inches under to 6 inches over the nominal length of the bundle. No piece shorter than 9 inches admitted, the percentages under 4 ft. referred to apply on total footage in any one shipment of the item. ¾ inch added to face length when measuring length of each piece. NESTED FLOORING is random length flooring bundled end to end continuously in 8 ft. long (nominal) bundles. OAK regular grade requirements apply.

PECAN 8-18 inch pieces will be admitted in 3/4x2¼" as follows: First Grade = 4 pcs. Second Grade = 8 pcs. Third Grade = as develops. Average Length: First Grade = 42 inches. Second Grade = 33 inches. Third Grade = 30 inches.
BEECH, BIRCH & HARD MAPLE 9-18 inch pieces will be admitted in 25/32"x2¼" or ¾"x2¼" as follows: First Grade = 4 pcs. Second & Better Grade = 8 pcs. Third Grade = 35 pcs.

4-7 *Hardwood flooring grades.* (Courtesy of the National Oak Flooring Manufacturers Association.)

variation and striking color contrasts. Number two common is a utilitarian floor that has more imperfections. It is suitable for cabins or rooms where appearance is not a primary concern. It will require more filling to repair defects than any of the other grades.

Plank and strip flooring

Plank and strip solid-wood flooring are similar in many respects. They differ, however, mostly in the widths of the two product categories. *Strip flooring* generally comes in widths of $3\frac{1}{4}$ in or less, while *plank flooring* includes boards that are 4 in wide or greater. The thickness of these floors can range from $\frac{5}{16}$ to $1\frac{1}{2}$ in, yet the most commonly used thickness is $\frac{3}{4}$ in. Both plank and strip flooring come *side- and end-matched*. Boards are said to be side-and-end-matched when there are tongues and grooves along the edges of the board as well as at the ends. Flooring products are available that either have no tongue and groove or are side-matched only, but they do not provide as strong a floor. When both the sides and ends are matched, the floor is locked in as a single unit. In addition, side-and-end-matched flooring can be blind-nailed through the tongue; boards that are not side-matched have to be face-nailed. This increases labor time and costs, since the holes must be filled prior to sanding and finishing. Furthermore, plank and strip flooring are also similar in that they both conform to the NOFMA grading rules shown in Fig. 4-7. This helps simplify and takes the confusion out of the selection process.

Bundles

Strip flooring is packaged together in bundles. There are three kinds of bundles: *nested bundles, average length,* and *specified length.* While average length and specified length are somewhat self-explanatory, nested bundles may require some explanation. A nested bundle comprises random-length boards bound together in 8-ft bales. It consists of four layers of boards that are three boards across, or a six-layer by two-board segment. The length of the boards should be from 9 to 102 in. No board shorter than 9 in will be used. Each bundle should equal 24 *board feet* of flooring. A *board foot* is a unit of measure for lumber in which the nominal dimensions are equivalent in volume to 1 ft (length) × 1 ft (wide) × 1 *in* (thick). Since these nested bundles come in relatively standard-size parcels, it makes estimating job costs and square footage much easier.

Measuring and estimating

An average 8-ft nested bundle is said to cover approximately 18 ft². When you figure any wood job, it is necessary to add an additional 5 to 6 percent to allow for cuts and *culls* (inferior or useless pieces). When an area has a lot of angles or the material is to be laid on a diagonal, it is advisable to add 10 percent or more to the original *net* (exact) square footage of the job. Calculating square footage in odd-shaped rooms is relatively easy after you have done it a few times. Remember to break the area down into small sections. Calculating square footage for wood is a great deal simpler than figuring square yardage for carpet. In carpet, you have to allow for the width of the roll and the placement of the seam. In wood flooring, it is just a matter of multiplying the length by the width of a given area. If the area has many nooks and crannies, just treat each area separately, then add the totals at the end (Fig. 4-8)

A typical square-footage takeoff will look like this (Fig. 4-9) (Please note that this takeoff formula can be used for any floor covering material that requires square-foot calculations. Those products may include vinyl floor tile, ceramic tile, wood parquet tiles, or even commercial carpet tiles.): Begin by drawing a floor plan sketch as described in Chap. 1. Then measure all the areas that will receive new flooring. Also include measurements for areas that will not receive flooring, such as existing cabinets, so they can be properly deducted later. Draw an X with a pencil in a given area, and assign it a sequential number. Place that number in a small circle at the center of the X. Begin a column, either on the same page or on a separate sheet of paper, to list the size of each numeric section. For example, section 1 of Fig. 4-9 shows the room to be 12 × 15; list it as (1)12 × 15. Be sure to deduct the area where the existing cabinets are installed; otherwise, the square footage will be too high. Note those deductions directly below section 1. Put those figures in angle brackets < > to signify that a deduction is to be taken. Continue to break down each room into separate little sections. Place an X in each section, and record it in the column. Alcoves and closets should be treated as individual units. After each room has been dissected, check your calculations for accuracy.

Multiply the length and width of the sections and extend them to the right-hand column. When all extensions have been completed, add the total square footage, including deductions. In Fig. 4-9, that total is 527.25 ft². (This is the *net square footage.*) As stated previously, *5 percent waste* must be added to allow for cuts and culls. Therefore, multiply 527.25 by 5 percent. That comes to 26.36 ft². Then add those

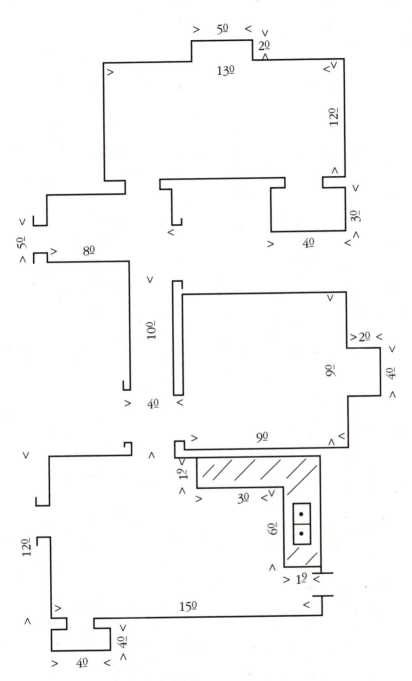

4-8 *Measure rooms for wood flooring.*

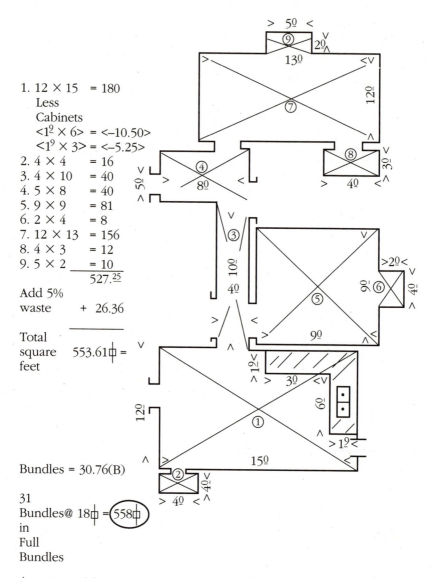

1. 12 × 15   = 180
   Less
   Cabinets
   <1² × 6> = <−10.50>
   <1⁹ × 3> = <−5.25>
2. 4 × 4   = 16
3. 4 × 10   = 40
4. 5 × 8   = 40
5. 9 × 9   = 81
6. 2 × 4   = 8
7. 12 × 13 = 156
8. 4 × 3   = 12
9. 5 × 2   = 10
                527.²⁵

Add 5%
waste   + 26.36
            _____
Total
square   553.61⌷ =
feet

Bundles = 30.76(B)

31
Bundles@ 18⌷ = (558⌷)
in
Full
Bundles

4-9 *A wood flooring takeoff.*

two figures to arrive at a total square footage necessary to properly complete the job. That amount is 553.61 ft². However, if the wood flooring you are purchasing comes in nested bundles, you have to calculate the square footage of the next-highest nested bundle. If this is not done, you will have to pay for more square footage than you calculated in your proposal.

Assuming that the average nested bundle is 18 ft², if you divide the total square footage (553.61) by 18, it equals 30.76 bundles. Since this is not an even number of bundles, raise the bundle amount to 31 and multiply by 18 ft². The total is now 558 ft². That is the number of square feet—and 31 is the number of bundles—you will order. (If you were calculating for another product, such as a vinyl tile, and the tile came 45 ft² per box, you would use 45 as your constant number instead of 18. Whenever a manufacturer will not break up product units, it is imperative to calculate to the next full unit.) Now that the square footage has been figured, refer to Chap. 1 for job cost calculations. The formula is the same; only the product and labor components will change slightly.

Trims and moldings

One of the main areas in which calculating job costs for wood flooring differs from those for other floor products is trims and moldings. Trims include such items as reducer strips, thresholds, and stair nosings. Moldings include wall base, quarter-round, or shoe molding. Since wood floors require an expansion gap at all walls and vertical rises, some type of molding is needed to hide the visible gap. If there is an existing molding, you must include in your estimate the time needed to remove and reset these items. If there is an existing wall base that should not be disturbed for some reason, a quarter-round molding can be added to the wall base to cover the expansion gap.

Trims will also be required in most wood flooring installations because of the relative thickness of the product. If left exposed without a suitable transition piece, the flooring would become a hazard and someone could trip and become injured. Reducer strips are necessary to provide a logical change from a thinner flooring, such as sheet vinyl, to a higher-profile material such as wood flooring.

Thresholds are also an interesting transition trim piece that separates one floor covering surface from another. The *baby threshold* is used in situations where a reducer strip would be impractical. For instance, if a wood floor is to be installed next to a carpeted area, a baby threshold can be used to provide an attractive transition (Fig. 4-10). Baby thresholds can also be used to lend a nice finished appearance at a sliding glass door or at a doorway. A *T molding* is a trim piece that separates two different flooring materials of the same height (Fig. 4-11). This provides an excellent transition from ceramic tile to hardwood flooring. It could also be used between woods of the same type or between woods of a different style. The T molding, as with all trim

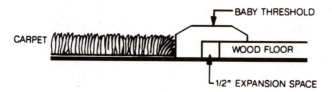

4-10 *A baby threshold.* (Courtesy of Bruce Hardwood Floors.)

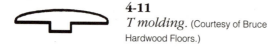

4-11

T molding. (Courtesy of Bruce Hardwood Floors.)

pieces, helps hide the rough, cut edges of the two surfaces and gives an appealing visual appearance to the transition area.

A *stair nosing* is a trim piece that finishes the front edge of a stair. Stair nosings are functional as well as aesthetically pleasing. Wood flooring cannot be laid right up to the edge of a stair, so stair nosings are needed to make the proper transition.

Trims and moldings give any job that *finished* look. They give a room the appearance of clean, straight lines. Attention to these details can make a job look spectacular. Be sure to include all these products in your cost estimate, because most customers assume that trims and moldings are included in the bid proposal. Wood finish pieces are very costly. If these products are overlooked in the estimate, they may have to be paid for by you. On the other hand, if you approach trims and moldings as a potential profit enhancer, you could help increase your gross revenue considerably. A 25 to 30 percent markup on transition pieces could lead to greater profits.

Job site inspections for solid strip or plank flooring installations

Job site inspections are important for all floor covering installations. Yet, they are particularly vital for wood flooring because of the volatility of the product itself. The wood's susceptibility to moisture gain or loss becomes increasingly evident once the floor is installed. An examination of existing conditions, and current floor coverings or subfloors, will help determine the appropriate preparation techniques needed to get the floor ready to receive the new material. Time, money, and energy can be saved by properly planning each individual installation.

The existing type of subfloor over which the wood flooring will be installed is the first major consideration. There are basically three kinds of subfloors: plywood, wooden plank, and concrete. Each of these substrates presents different challenges for a solid-wood floor installation. For new construction with subfloors made of plywood, $\frac{5}{8}$- or $\frac{3}{4}$-in performance-rated products should be used. A $\frac{3}{4}$-in OSB can also be used as an appropriate substrate. These boards should be installed at right angles to the joists and nailed every 6 in. The proper spacing of $\frac{1}{8}$ in at the panel ends and $\frac{1}{4}$ in at the panel edges is desirable. Accurate nailing is necessary to keep the boards solid, since a solid subfloor prevents floors from squeaking.

If the plywood floor is an existing subfloor that has been in place for some time, check whether it is level and flat. Ideally, a subfloor should be flat to $\frac{3}{16}$ inches in a 10-ft span. Use a 6-ft level to check the floor in several areas for any irregularities. A subfloor that is not flat is more problematic than a floor that is not level. A subfloor that is not flat can cause *overwood,* where two adjoining strips of wood have different heights. It can also cause creaking, looseness, or open joints. If *hills* or *valleys* exist in a subfloor, sand down any hills and fill any valleys with a high-compression underlayment patch or self-leveling cement compound. Where the variations in the subfloor are severe, it may be necessary to drive shims between the joists and the subfloor if the joists are accessible.

Job inspection for wooden plank subflooring and plywood flooring is similar. Wooden plank subflooring should be checked to make sure all planks are properly nailed and that there are no loose boards. Use 8d nails or screws to secure the boards into the joists. If low spots are found in the subfloor, raise them to an acceptable level; otherwise, squeaks will develop.

Another important condition to check in the job inspection is the moisture content of the plywood or wooden plank subfloor. This can be done by using a good-quality moisture meter. The moisture content of these wood subfloors should not exceed 14 percent. Since this percentage could exceed the moisture content of the wood flooring (which should be between 6 and 9 percent with a 5 percent allowance for pieces outside that range), it could cause a solid oak floor to *cup* (the edges of the flooring strips raised due to excessive underside moisture). If you find high moisture readings in the wood subfloor itself, check for spills, leaks, improper landscape drainage, or poor crawl space ventilation.

When doing your normal walk-through job inspection, you must check the basement or crawl space for excessive moisture. Thoroughly inspect all areas to see that there is proper ventilation through vents or other openings. Basements must be dry and free of any condensation. Buildings without basements should be insulated against ground moisture by a 6-mil-thick polyethylene film. This moisture barrier (vapor retarder) should be laid over all the crawl space area to prevent moisture from saturating the subfloor and joists.

The polyethylene film should also be extended up the wall in the crawl space at least 4 to 6 in. The polyethylene film can be secured against the wall with bricks, cinder blocks, or sand. This will keep the entire vapor barrier in place and anchored. Regularly inspect crawl spaces after heavy rains to make sure that no water has accumulated on top of the plastic groundcover. This could negate the effect of the vapor barrier altogether if the accumulation is substantial.

Concrete subfloors

For concrete subfloors, the slab must be checked for moisture. Refer to Chap. 1 for the proper methods for checking concrete slabs for moisture content. It is imperative that concrete subfloors be dry and well cured to a nonpowdery finish. If a subfloor shows signs of excessive moisture, allow it to dry naturally. If time is of the essence, drying can be hastened with ventilation and heat application.

A concrete floor must also be checked for flatness (hills or valleys). Just as with a wood subfloor, the subfloor needs to be flat; so grind down any hills and fill any valleys. The subfloor should also be evaluated for levelness. A concrete slab that is not level may need to be made so with the use of a self-leveling compound. Contact a reputable manufacturer of leveling products and follow instructions accurately.

Moisture barriers

There is always a possibility of some moisture's migrating through the concrete. To protect the wood flooring, an appropriate vapor retard is placed directly on top of the slab by either of two methods: a 4- to 6-mil polyethylene film or two layers of 15-lb asphalt felt or building paper and mastic. (A vapor barrier is also recommended over a wooden plank subfloor.) How and where these vapor barriers will be placed will depend upon the type of installation method chosen for the new finished wood flooring. Both systems of installing the vapor barrier with slab construction—the polyethylene and asphalt felt or building paper system—are secured by using cold cutback asphaltic mastic. The asphalt felt or building paper system uses two layers of material glued on top of each other (Fig. 4-12). The first layer is glued to the

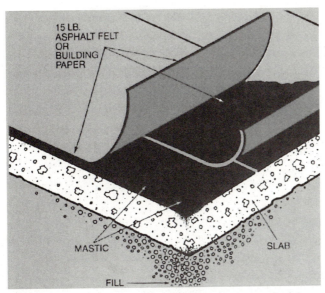

4-12 *Moisture retarder using two layers of asphaltic felt or building paper.* (Courtesy of the National Oak Flooring Manufacturers Association.)

slab, and the second layer is glued to the first layer. The edges of each layer are overlapped. The seams of the second layer are staggered at least 6 in away from the seams of the first layer. Let the mastic set up at least 2 h before you lay in each layer of felt or paper. The polyethylene method requires only one layer of film because it is more nonporous than the asphalt felt or building paper. The film should be allowed to go under the baseboards. The mastic should be spread and allowed to set up for $1\frac{1}{2}$ h before unrolling the film onto it. The film may then be rolled with a linoleum roller or walked on to make certain that a suitable bond has been made.

Method for installing solid strip or plank flooring over a concrete slab

There are two types of installation systems for installing solid strip or plank flooring over a concrete slab: the plywood-on-slab system or the screeds system. The plywood-on-slab method (Fig. 4-13) uses a minimum $\frac{3}{4}$-in-thick exterior-grade plywood over one of the vapor barrier systems recommended for concrete floors. The plywood sheets should be laid diagonal to the way the wood floor will be laid and secured with a power-actuated concrete nailer or by regular hand-nailed concrete nails. The power-actuated nailer is much faster and easier than

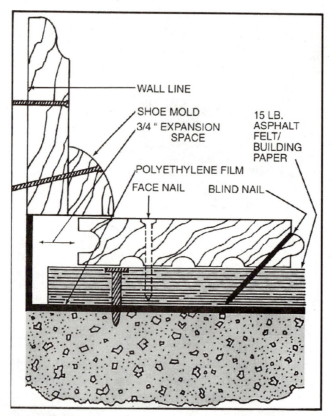

WALL LINE

SHOE MOLD

3/4 " EXPANSION SPACE

15 LB. ASPHALT FELT/ BUILDING PAPER

POLYETHYLENE FILM

FACE NAIL   BLIND NAIL

4-13 *Plywood-on-slab method of installing strip oak flooring.* (Courtesy of the National Oak Flooring Manufacturers Association.)

nailing by hand. The underlayment plywood can also be glued in place by asphaltic cutback adhesive. The plywood should be cut into 4 × 4 sheets. The back of the plywood should be scored $\frac{3}{8}$ in deep on 12-in × 12-in grid. Allow the mastic to set up for at least 12 h before you lay the plywood in place.

The screeds system is a common method for laying a strip or plank wood flooring (less than 4 in) over a concrete slab (Fig. 4-14). *Screeds* are preservative-treated 2-in × 4-in pieces of wood that are secured to the concrete slab. These screeds provide a nailing base and are placed at right angles to the finished floor. They are installed on their flat surface, at 12 in on center, in a bed of heavy mastic. After the mastic has dried, a layer of 4- to 6-mil polyethylene film is laid over the screeds. This film acts as a moisture barrier. The wood flooring is then nailed to the screeds through the polyethylene film. If a plank 4 in and wider is desired, the plywood-on-slab system should be used to ensure satisfactory nailing capabilities.

Method for installing solid strip or plank flooring over a wood joist floor system

The most common and best way to install solid strip or plank flooring is onto a wood joist floor system. Unless there are severe water conditions, this limits the likelihood of adverse moisture problems considerably. It provides an excellent surface on which to nail the new finished floor.

Whether the subfloor is made of plywood or wooden plank, it should be overlaid with 15-lb asphalt felt or building paper (Fig. 4-15). This acts as a good moisture retarder. The polyethylene film that is used over concrete subfloors is not recommended over wood subfloors because it limits airflow and natural moisture evaporation. It can create condensation which in turn could cause the finished floor to cup. The felt or paper should be stapled to the subfloor to keep it from moving during installation. The use of this felt or paper will also help to properly seat the finish floor nails, lessen squeaks, keep out insects

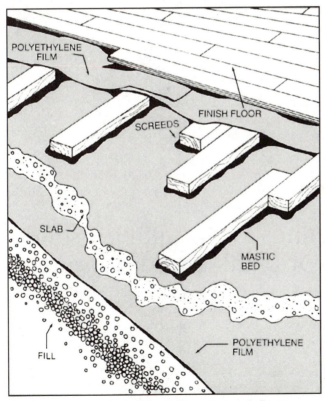

4-14 *Screeds method of installing strip oak flooring on slab.* (Courtesy of the National Oak Flooring Manufacturers Association.)

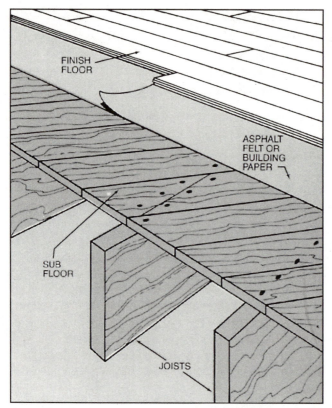

4-15 *Wood joist construction using square-edge board subfloor.* (Courtesy of the National Oak Flooring Manufacturers Association.)

and dust, and create a clean, suitable work area to install the new floor. A nail-down wood floor should not be installed over particleboard.

Guidelines for installing a solid strip or plank wood flooring

The following guidelines are meant to be general instructions for installing solid wood strip or plank flooring. Consult with the wood manufacturer for specific installation procedures for a product.

The wood flooring should be delivered to the job 4 to 5 days prior to installation. This should allow sufficient time for the wood to stabilize and acclimate to the new environment; however, more time may be necessary. The flooring should not be unloaded in the rain or dense fog, as this could increase its moisture content. The wood flooring should not be delivered to the job site until after all other materials that were installed with water have had time to thoroughly dry. These materials include sheetrock (drywall), concrete, and masonry. Since these products release moisture as they cure, they could adversely affect the moisture level of the indoor environment and of the wood flooring. Before installation, all doors and windows should be in place and airtight. Temperature and humidity levels should be similar to those at the time of occupancy. The heat should be turned on during cold weather, and proper ventilation must be maintained during warm weather. A consistent room temperature of 60 to 70°F must be maintained from the time the wood is delivered to the job site until well after the building is occupied. Thereafter, an appropriate environment must be maintained to prevent unfavorable moisture conditions from developing. If the wood is not allowed to properly acclimate and stabilize its moisture content, cracks could develop between the boards. When installed at too high of a moisture level, the wood will subsequently shrink and cause gaps between the strips of flooring. But if the moisture content is too low at the time the floor is installed, when the wood regains its normal moisture content, it will cause the floor to expand and buckle.

Once you are sure that every precaution has been taken to stabilize moisture levels of the flooring and environment, it is time to make a plan for laying the floor. (Since the installations of strip and plank flooring are very similar, the term *strip flooring* is used to mean both strip and plank floors that are ¾ in thick.)

Design and layout

Transforming a dingy wooden plank, concrete, or plywood subfloor to a sparkling-clean strip flooring surface must begin in the design and layout stage. Since most walls are not always square and nearly all subfloors are not perfect, some adjustments will need to be made. These adjustments are best made before any boards are laid. The *five P's* are certainly appropriate for laying a wood floor: *proper planning prevents poor performance.* The better the plan, the greater the performance.

If all the preparation work has been done properly, the actual laying of the floor should be easy. Since nearly every room has an area to which the eye is inevitably drawn, this center of attention should be as nearly perfect as possible. When you design a layout, the starting line is the most critical choice you have to make.

Two possible choices for starting lines are the traditional starting line and the center starting line (Fig. 4-16). The traditional starting line begins at one wall and works in one direction to the opposite wall. The center starting line begins in the middle of either the room or an important viewing area such as a hall or doorway. From that center

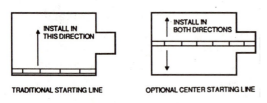

4-16 *Starting lines.* (Courtesy of Bruce Hardwood Floors.)

starting line, the boards are laid in both directions to the opposite walls. Regardless of the method chosen, the laying of the first few rows (or *courses*) is the most important in any flooring installation.

Laying the floor

Select the straightest and longest boards possible for the first several rows. These first courses, as well as the last few courses at the opposite wall, will be *free-nailed* (nailed directly into the face of and through the board and then *countersunk*) (Fig. 4-17). The "field" (middle-of-the-room) rows are *blind-nailed* (nails are driven at a 45° angle through the strip where the tongue leaves the shoulder) (Fig. 4-18). Machine-driven fasteners help accomplish this chore easily and efficiently (Fig. 4-19). Before you nail these boards in place, mark the position of the joists on both the starting and finishing perimeter walls. A floor should be laid where the strip flooring is installed perpendicular to the floor joist. It is necessary for good installation to have the beginning and ending rows nailed securely into the floor joists.

The starter line is established by driving a nail $\frac{3}{4}$ in plus the width of the floor $2\frac{1}{4}$ in from the end wall close to the corner of the starting wall. Do the same thing at the opposite corner, and attach a string line to both nails (Fig. 4-20). Face-nail the first board next to, and touching, that line. Once the first row of flooring is face-nailed, it is a good idea to blind-nail it to strengthen the bond. As discussed earlier, make sure you have allowed for the proper expansion gap at the starting wall. If, for some reason, the wood has to be laid close to the wall, undercut the drywall to provide for the expansion and contraction underneath it. The first course should be laid with the groove to the wall and the tongue toward the direction you are laying the floor.

Racking the floor

Begin *racking* the floor by laying out six or seven rows of boards so nailing can be accomplished in quick succession. The boards should be laid out with the end joists staggered at least 6 in apart. Racking provides a kind of minipreview of how the floor is going to look, section by section. It is your responsibility to take a few moments at this time to harmonize all the various grains, colors, and lengths of board into an appealing presentation. Work from several different bundles or boxes of flooring to guarantee a good mixture of varying board lengths and color shades. This prevents the possibility of having a wildly off-color bundle in one section of the floor destroy the natural flow of the shades. By staggering the end joints you stop the boards from looking

4-17 *Face-nailing machine.* (Courtesy of Stanley-Bostitch.)

4-18 *Blind-nailing a strip wood floor.* (Courtesy of the National Oak Flooring Manufacturers Association.)

4-19 *Blind-nailing machine.* (Courtesy of Stanley-Bostitch.)

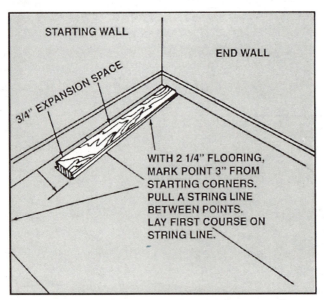

4-20 *Starter line, first course of a strip floor installation.*
(Courtesy of the National Oak Flooring Manufacturers Association.)

too symmetric. Staggered joints are more pleasing to the eye and give the room greater visual depth. Bundles that have short pieces will generally contain a significant number of them. Use those pieces wisely by spreading them evenly throughout the floor. Use these pieces in the field whenever possible.

The soundness of the finished floor is determined in the racking stage. It is here that you should check for obvious defects. Saw and planer dips, loose knots, splits, and splintered ends are some of the imperfections to watch for. Where a portion of a board has a defect, cut out the bad part and use the remaining good piece. It isn't necessary to discard all the board.

Nailing the floor

Once the first few starting rows are securely in place and several other rows are properly racked and ready to go, it is time to start nailing the floor. NOFMA has prepared a nailing schedule (Fig. 4-21) as a guideline for accurate nailing. It shows the type and size of nail to be used for any given floor. It specifies the spacing between each nail and the minimum number of nails required for each board—which is 2. It also explains the minimum requirement for kinds of subfloors that can be used in certain situations. Study it carefully before you begin nailing, to avoid complications.

After several courses have been laid, sight down the rows to check for straightness. Any waviness should be corrected before you continue with more courses. If the waves cannot be altered, remove the bad pieces as necessary. As far as tightness of the boards is concerned, an occasional hairline gap between the side matches is acceptable. The end should abut with an allowance for slight off-squareness.

Almost all floors require some fitting around vertical obstructions. Cabinets, stairs, borders, and heat registers are just a few of the many impediments to flooring. *Ripping* is a term used to denote cutting or trimming a board to fit along its length. When you are working around openings such as heater vents, trial-fit boards around the open area. Decide whether the groove or the tongue will be saved, mark accordingly, then cut. A visible focal point, such as a fireplace, should be framed in wood to give the floor a more finished appearance. A hearth border is easy to do, yet has a pleasing effect. Using a miter box to achieve precise 45° angle cuts, trim pieces to fit around the fireplace. Work the flooring pieces around the frame so that the tongue and grooves are engaged. After you have cut around or placed trim pieces around all the obstructions, it is time to complete the installation portion.

NAILING SCHEDULE

NOFMA Hardwood Flooring Must be installed over a Proper Subfloor
Tongue & Grooved **MUST be Blind Nailed**
Square Edge must be Face Nailed
Inadequate nailing contributes to cracks and noisy
floors by allowing movement of the flooring.

A slab with screeds 12" o.c. does not always require a subfloor.

STRIP T & G Size Flooring	Size Nail to be Used	Blind Nail Spacing along the length of strips. Minimum 2 nails per piece near the ends. (1"-3")
$^3/_4$ x 1$^1/_2$", 2$^1/_4$" thru 3$^1/_4$"	2" serrated edge barbed fastener, 7d or 8d screw or cut nail, 2" 15 gauge staples with$^1/_2$" crown. On slab with $^3/_4$" plywood subfloor use 1$^1/_2$" barbed fastener or staple.	In addition - 10-12" apart - 8-10" preferred.

MUST Install on a Subfloor

PLANK $^3/_4$ x 4" to 8"	2" serrated edge barbed fastener, 7d or 8d screw or cut nail, 2" 15 gauge staples with $^1/_2$" crown. On slab with $^3/_4$" plywood subfloor use 1$^1/_2$" barbed fastener or staple.	8" apart

FOLLOW Manufacturer's Instructions for Installing Plank Flooring.

Widths 4" and over must be installed on a Subfloor of $^5/_8$" or thicker plywood or $^3/_4$" boards. On slab use $^3/_4$" or thicker plywood.

$^1/_2$ x 1$^1/_2$" & 2"	1$^1/_2$" serrated edge barbed fastener, 5d screw, cut steel, or wire casing nail.	10" apart $^1/_2$" flooring must be installed over a MINIMUM $^5/_8$" thick plywood subfloor.
$^3/_8$ x 1$^1/_2$" & 2"	1$^1/_4$" serrated edge barbed fastener, 4d bright wire casing nail.	8" apart

SQUARE-EDGE FLOORING (Not Tongue and Groove)		
$^5/_{16}$ x 1$^1/_2$" & 2"	1" 15 gauge fully barbed flooring brad.	2 nails every 7"
$^5/_{16}$ x 1$^1/_3$"	1" 15 gauge fully barbed flooring brad.	1 nail every 5" on alternate sides of strip.

For additional information - write to:
National Oak Flooring Manufacturers Ass'n.
P. O. Box 3009, Memphis, TN 38173-0009

® CERTIFIED FLOORING GRADE
NOFMA trademark assures quality.

4-21 *Nailing schedule.* (Courtesy of the National Oak Flooring Manufacturers Association.)

The final rows

As you approach the opposite wall, check the distance between the face of the board and the wall. Do this at a number of locations. The rows should be parallel to the wall. If they are more than $\frac{3}{4}$ in out of parallel, then the boards should be tapered. The amount of distance that needs to be gained and the number of rows left will determine the size of the tapering. Boards can be tapered by striking a line for the entire length of a grooved edge and trimming it with a jointer or hand plane. It is advisable to do as much tapering in the middle of the room as possible because once the wall base or shoe molding is installed, it will highlight lines that are not parallel. Proceed toward the far wall where the last few rows will be face-nailed, just as the first several rows were. These rows can be pried into position by placing a pry-bar against a scrap piece of flooring at the wall line. The pry-bar can be turned to force the last board back into the previous one. Hold the pry-bar tightly in place with your foot as you face-nail the final row. Be sure to leave the same expansion gap as at the beginning row.

Now that the floor is completely installed, it must sit for a minimum of 1 week—or up to 1 month—to acclimate to the structure. The floor will expand and contract during this time in relation to the atmospheric environment. If the sanding is done too soon, additional changes may take place in the wood that will require further sanding or filling. Give the floor enough time to relax.

If the floor must be used during this waiting period, protect it with cardboard or heavy paper. Remember that spills present the greatest danger to a floor during this period. Certain products such as paint, solvents, water, or lacquers could stain the floor and leave an unsightly mess that will be nearly impossible to remove. Check the floor periodically to observe the changes taking place and to ensure that the newly laid floor is being properly protected.

Nailing wood flooring to screeds

When screeds are applied directly to a concrete slab, there is no subfloor. It is therefore imperative to make certain that the flooring is properly nailed to the screeds. When a board goes over a staggered screed, nail to both screeds. When a board covers a continuous screed, nail at all intersections. Take care not to allow neighboring boards to have end joints that stop over the same empty space between screeds. This could weaken the floor and cause springiness. Squeaks could then develop in the floor.

Installation of plank flooring

Plank flooring is installed using the same techniques as strip flooring, except for some minor differences. Plank flooring consists of boards that are 4 in wide and above. A typical random-width floor has boards that are 3, 5, and 7 in wide. This pattern will repeat until the far wall is reached. Nail plank flooring according to the nailing schedule in Fig. 4-21. This table shows that the flooring should be nailed every 8 in.

One of the main differences between strip and plank flooring is that the wider plank boards are more susceptible to changes in moisture and humidity. This causes the planks to move more. In addition, the wider boards have fewer nails per square foot of coverage area than the narrower strip flooring. Consequently, plank flooring has a greater likelihood of cupping and squeaking. Therefore, plank flooring should be properly acclimatized at all stages of the installation.

Planks that are 4 in or wider should be face-nailed or screwed to the joists and subfloor. Round holes are bored into the plank flooring at each end of each plank as well as at intervals along the plank. Into these holes are driven drywall screws (Fig. 4-22). The length of the screw can vary between 1 and 2 in or more depending on the type of subfloor. Once these screws are countersunk below the surface of the planks, the holes can be covered with wood plugs that are glued in place.

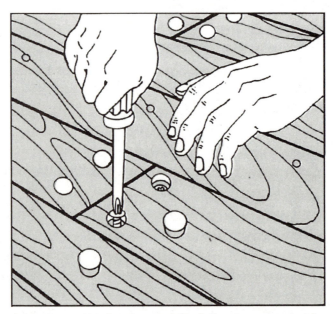

4-22 *Countersink screws in plank flooring, cover with plugs.* (Courtesy of the National Oak Flooring Manufacturers Association.)

The screws provide functional stability to a plank floor, while the wood plugs create a pleasing accent to an otherwise mundane-looking environment. A cabin or vacation home can be made to appear quite rustic whenever wood plugs are used. The use of contrasting-color plugs from different wood species provides an attractive alternative to same-species plugs. Be careful when you lay out the floor not to use so many screws that the surface becomes permeated with them. Planks that are wide and long should be screwed at the butt ends as well as at every second or third joist. Aim to create an appealing pattern that does not look too overwhelming.

Sanding and finishing a wood floor

Now that the floor has been properly installed, it is time to proceed with the steps that enhance the wood's appearance: sanding and finishing. The sanding process will remove any *overwood,* or dirt and minor scratches, that occurs during installation. The finishing procedure not only will add color and enhance the look of the grain, but also will provide a protective layer to the floor so that the surface can sustain its beauty for many years to come.

Once the flooring is installed, it should be allowed to acclimate to its environment for at least 1 to 3 weeks. During this time, the floor expands and contracts according to climatic changes. After the floor has adjusted to its surroundings, minor cracks, raised edges, and wedged ends can be sanded, filled, and finished to give the floor a pleasant appearance.

Sanding should not be started until all other work inside the structure has been completed. All other trades should be finished except for the final coat of paint on the wall base moldings. Since the sanding process is extremely messy, care should be taken to protect personal items as well as the surrounding interior surfaces of the building. The sanding and finishing phase will create a great deal of noise, odor, and dust. The dust could be everywhere and on everything. Prepare the owner of the building for this eventuality. Ask that books, knickknacks, and all other objects be removed from not only the rooms to be sanded, but also any adjacent rooms. The dust tends to migrate to other areas whenever possible. Because the dust is of such a fine quality, all doors or openings should be sealed. Even the bottoms of doors should be protected against dust getting under them by placing towels behind the thresholds of every interior door. In addition, kitchen cabinets can be sealed with masking tape to keep out the dust. Sanding machines are now on the market that capture a great amount of wood dust. These products should be used when-ever practical.

The floor should be swept clean, and the surface should be thor-oughly inspected for any protruding nail heads. All boards that have been face-nailed should have the nails countersunk below the surface of the wood at least $\frac{1}{8}$ in. Any nails that do not look deep enough should be reset beneath the top of the board with a *nail set* (a fine-point tool struck with a hammer to countersink a nail). Be careful to check around the perimeter of the room where face nailing was most likely. The importance of setting nails cannot be overlooked. If an elec-tric sander comes in contact with a protruding nail head, it can produce sparks which can create a fire hazard in the dust bag of the sander.

After you check for exposed nail heads, examine the floor for cracks and gaps. Nearly every floor will require some filling with a commercial wood filler to cover nail holes or obvious irregularities. Filling gives a wood floor a more uniform appearance. (Carefully read all instructions from the manufacturer of the filling material to be sure to apply the product according to specifications.) Spot filling and grain filling even out the face of the board. They sand easily and stain similarly to wood.

Equipment

The three machines needed for sanding and finishing a wood floor are a drum sander, an edger, and a buffer. The drum sander does the majority of the work. It levels the main area of the floor, thus remov-ing any overwood or high spots. The edger sands those areas that the large, bulky drum sander cannot. The edger can get close to the walls with its rotating disks (Fig. 4-23). The buffer can be used in both the sanding and the finishing stages. In the sanding phase, it can blend the raw surface before staining and finishing. In the finishing stage, the buffer can be employed to rough up a dried finish coat to get it ready to receive another coat of floor finish.

In addition to the machines that are necessary, the following tools and materials are needed: coarse-, medium-, and fine-grit sanding pa-per; belts and edger disks; nails; 6d and 8d case nails; broom; ham-mer; vacuum; nail puller; paint scraper; and hand sanding block, to name just a few.

Sanding a new strip floor

The drum sander does the bulk of the sanding in the main field area of a room. This machine removes a thin layer of wood from the sur-face of the floor. Sandpaper is wrapped around a metal rotating drum

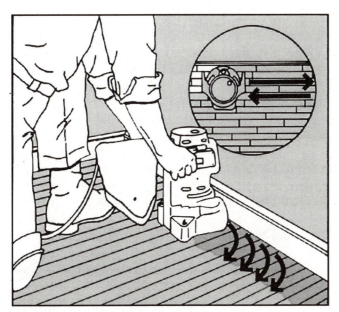

4-23 *The power edger.* (Courtesy of the National Oak Flooring Manufacturers Association.)

that is sleeved with hard rubber. This rubber sleeve provides a cushion against the floor and supplies traction for the sandpaper.

The most important feature of the drum sander is the method used to raise and lower the drum. There are two types of drum-elevating mechanism: self-raising/self-lowering and lever-actuated. With the former type, the operator raises or lowers the handles of the machine manually which in turn lifts the drum off the floor; with the latter type, a lever is used to raise or lower the drum accordingly. The lever-actuated drum sander is the preferred model because it provides greater control during sanding. To begin, load the machine with coarse- to medium-grit sandpaper for the first sanding operation. Follow all manufacturer's instructions for proper loading procedures.

When you sand a strip or plank wood floor, you move the sander back and forth (with the grain) to remove a fine layer of wood with both the forward and backward passes. Place the machine along the right-hand wall about two-thirds of the way into the room. The machine is turned on with the drum raised off the floor. As you start each pass, gently ease the drum onto the floor from a walking position. When you approach the opposite wall, slowly begin raising the drum off the floor until it is no longer in contact with the surface. The raising and lowering of the drum at the beginning and end of each pass helps *feather* the face of the wood and create a smoother floor.

This feathering disguises the individual start-and-stop marks of each pass. The machine must always be moving when the drum is either raised from or lowered to the floor; otherwise, a hollow (or stop mark), in the shape of the drum, will be left in the floor. This hollow will not be easy to remove, so keep the sander moving whenever it is in contact with the floor (Fig. 4-24).

After you complete the forward pass, ease the drum gently down to the floor to begin the backward pass. Sand the exact same path as the forward pass by pulling the machine backward. As you reach the beginning point, raise the drum to feather the surface of the wood. Then move the machine slightly to the left (approximately 3 to 4 in). Sand the next pass just as the first pass. The second pass will overlap the first pass, which gives an even appearance to the surface. All passes are sanded in this fashion until two-thirds of the room is completed. At that point, turn the machine around 180°, overlap the previously sanded passes by 2 or 3 ft, and then sand the remaining one-third of the room in the same manner as the original passes (Fig. 4-25).

Once the entire main area has been sanded for the first time (the first *cut*), use the edger to sand all areas not covered by the drum sander. The grit of the sandpaper used on the edger is determined by

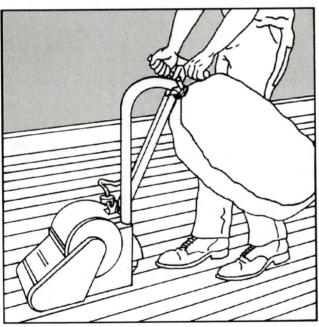

4-24 *Keep sanding drum in motion whenever it touches the ground.* (Courtesy of the National Oak Flooring Manufacturers Association.)

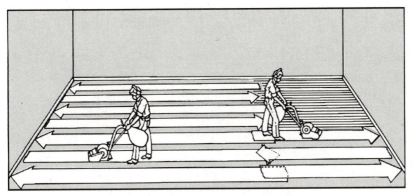

4-25 *Sanding directions.* (Courtesy of the National Oak Flooring Manufacturers Association.)

the grit used on the drum sander. When you sand a new floor, if coarse grit is used for the first cut, use medium grit on the edger. Similarly, if medium grit was first used, the fine grit will be acceptable.

The edger is a small, hand-operated sanding machine. It can sand in tight areas that larger sanders cannot reach. There are several types of edgers, but the most commonly used kind consists of a motorized sanding disk that is approximately 5 to 7 inches in diameter. The angle and depth of the cut can be adjusted by two wheels located inside the housing. A circular sheet of sandpaper is attached to a rubberized spinner disk that is set at a slight angle to the surface of the floor. The edger is a powerful piece of machinery. Per square foot, it can outsand any other floor-sanding tool. Inspect the floor's surface after the first cut to see if the drum sander and edger have provided an acceptable surface. If so, further sanding cuts will simply remove any minor scratches or irregularities left by the first cut.

Once the first cut is complete, repeat the process, using the next grit sandpaper (i.e., coarse to medium grit). Then repeat the process a final time, using fine-grit sandpaper. Conclude the sanding operation by hand-sanding and hand-scraping all peripheral edges, corners, and door casings and around any vertical obstructions (Fig. 4-26). It is important to sand in the direction of the grain of the wood, with even pressure. The same grit sandpaper selected for the last cut should be used for this procedure.

Next, the buffing machine is used to help blend the entire floor into one smooth surface. Attach either a screen disk or a sanding disk to the buffer, and work in the direction of the grain. This final, fine abrading of the wood is indispensable to achieve a truly admirable-looking surface.

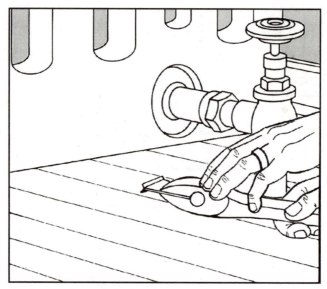

4-26 *Hand-scraping hard-to-reach places.* (Courtesy of the National Oak Flooring Manufacturers Association.)

The sanding procedure outlined above, which includes three sanding cuts and a final disk or screening step, is the recommended technique for achieving the best possible results. Eliminating one of the sanding cuts may give the floor a more rough-hewn surface that could prove to be unacceptable.

Every cut made with the sanding machine will produce some sanding marks. The goal is to make those marks less obvious by using finer and finer grits of sandpaper with each additional cut. Careful attention should be paid to feather one cut into the next.

Filling a wood floor

For best results, once a wood floor is laid, it should have all knot holes, cracks, and nail holes filled with a commercial, ready-made filler. Whether to trowel fill the entire floor after the first cut or spot-fill between finish coats will depend upon the personal style of the installer. It is important to check with the manufacturer of the filler that it is compatible with the floor finishes to be used.

Sweep and vacuum the floor thoroughly after the first cut, before you begin to fill. The filler is forced into the holes and cracks with a putty knife or trowel, and it is allowed to dry for 1 or 2 h.

Many current premixed products are nontoxic, water-resistant, weather-resistant, and nonflammable. They can be sanded, stained, painted, and nailed into. They are also safe to use with water- or solvent-based finishes. These premixed fillers are color-coded to match many of the common species of wood. They come in natural oak, walnut, maple, cherry, and mahogany, to name just a few. Custom shades can be created by combining two or more colors.

When you are filling the entire floor at once, mix a sizable amount of filler into a big glob, and begin spreading it from one corner of the room into the center. Do not be too concerned if a few spots are missed during this initial filling, because there will be other chances to fill them later between finish coats. When the entire floor has been filled, remove all filler from the surface of the wood.

Sand the floor again, using the same pattern as for the first cut. This sanding will smooth out the areas where any excess filler is slightly raised above the surface. Then sweep, vacuum, and inspect the floor for any spots that require immediate attention. Fill and sand these areas as necessary.

Final sanding of floor

Finish-sand with a fine-grit sandpaper to remove any scratches left by prior sandings. Use the drum sander on the main areas and the edger on the perimeter areas as before. Scrape and hand-sand all areas not covered by the machines.

The last step is to buff the floor with a screen-back disk attached to the buffer machine. Buffing once (with the grain of the floor) should provide a suitable surface. Sweep and vacuum one more time to remove any remaining dust and debris. Once this is done, it is time to apply the finish coats.

Finishing the floor

If the finishing process is not begun soon after the sanding has been done, it may cause the grain of the wood to become raised. If this happens, the surface of the wood will be rougher than it should be and so may require additional sanding.

Finishing a wood floor involves applying a stain, if preferred, and a coating that protects the surface from moisture and spills. If left untreated, wood flooring could quickly absorb many of the materials it comes in contact with, such as water, grease, oil, and dirt. Some of these substances are nearly impossible to remove because they penetrate deeply into the surface of the wood.

When you apply a wood finish, note that many of the most common problems associated with finishes are application related. The

person who actually applies the products to the floor is ultimately responsible for the performance of the product. Therefore, it is incumbent upon the installer to become thoroughly familiar with the
manufacturer's instructions. The single most recurring problem within
the industry is due to not properly following the manufacturer's directions. Some of the other, more common problems arise from not
letting the finishes adequately dry, improper floor preparation, and
mixing or altering the manufactured finish products. Mixing one or
more different manufacturers' products or adding substances such as
lacquer thinner to the finish can result in a major catastrophe. Each
individual manufacturer spends thousands of dollars on research and
product development. The time involved in bringing a product from
concept to marketing is enormous. These companies have tested
their products and have come up with the appropriate techniques.
The procedures that are acceptable for one manufacturer may not be
suitable for another. If you have any doubt as to how to use a certain
product, contact the manufacturer.

Before you begin floor finishing, follow the manufacturer's guidelines regarding safety and health precautions. If required, turn off any
pilot lights or ignition sources. Explosions or fires can be caused if the
instructions are not adhered to regarding flammability. Also, wear an
approved respirator, rubber gloves, and a pair of goggles as protection
against chemicals at this critical stage of the floor installation process.

Staining the floor

Whenever a color other than the natural shade of the wood is desired, the floor can be stained to a more pleasing tone. Stains can be
used in combination with a penetrating sealer to both color the floor
and strengthen the surface of the wood. The sealer is absorbed into
the pores of the wood, making it more durable and resistant to stains.
Allow the stain or sealer to thoroughly dry for at least 1 or 2 days. Although stains and sealers can be applied separately, the combination
products are preferable because they are easier and quicker to apply.

Stains are identified by two primary features: coloration process
and carrying vehicle. Stains obtain their color from dyes, pigments, or
both. The dye and pigment processes treat the wood in entirely different manners. In dye stains, the color becomes dissolved in the carrying vehicle. The dye is absorbed into the wood grain itself and
literally colors the wood. In the pigment process, the pigment is held
in suspension within the carrying vehicle, which in turn does not allow the pigment to be absorbed into the wood. The wood is not actually colored, as with the dye stains. Consequently, finishes applied

over dye stains are more durable than finishes applied over pigment stains, because the lack of adhesion with the pigment stains into the wood creates handling problems when the finish is applied. More care must be taken beforehand to ensure compatibility between the stain and the finish.

Stains can be transported in one of three vehicles: oil-based, fast-drying, and water-based products. Oil-based stains dry slowly, but do not raise the grain of the wood as much as water-based stains. Fast-drying stains are similar to oil-based stains except they contain dryers that expedite the curing process. Water-based stains cure faster than oil-based stains, are more environmentally friendly, and give off less odor.

The stain should be dry before you apply the finish coat. Stains can be applied with a lamb's wool applicator, a brush, or by hand-rubbing and wiping with rags. If rags are used, they should be of the absorbent, lint–free, cotton type. When the stain is applied, the excess that is not readily absorbed into the pores of the wood should be wiped up immediately; otherwise, it becomes too tacky to remove once it sets up. Make every effort to rub the stain deeply into the grain of the wood. Keep a piece of sandpaper nearby to quickly remove any sweat droplets, handprints, or watermarks. These marks could become magnified if left unattended. Picture-frame the room with a 3-in-wide band of stain to eliminate the potential for splashing it up onto the wall moldings. Thoroughly soak a clean rag with stain, then wring it out. Rub the stain into the floor in a half-moon, circular motion. Be sure to overlap previously stained areas to avoid any lap marks. Continue staining the entire floor, going in the direction of the grain.

The absorbent qualities of each species of wood are quite different. When a stain is applied to a raw piece of wood, the porosity of the grain (i.e., the earlywood and the latewood) determines the effect of the stain. For example, in a piece of Douglas fir, the earlywood is lighter than the latewood, but more porous. It therefore will absorb more stain and will take a darker finish. The appearance of a stained piece is the opposite of the appearance of an unstained portion (Fig. 4-27). Conversely, in red oak, where the earlywood pores are much larger and appear darker than the latewood pores, the effect is merely intensified. The latewood stays light-colored while the earlywood intensifies in darkness as the stain penetrates the pores.

After the first coat has been applied and allowed to completely dry, buff the floor with no. 1 steel wool or a fiber buffing pad on a floor buffer. Then clean and vacuum the floor to remove any dust or foreign material.

2—Earlywood of Douglas fir, far left, is lighter in color than latewood, but more porous. Thus it retains stain more readily and finishes darker in color. Large earlywood pores of red oak, left, already appear darker than latewood, and the effect is accentuated as stain accumulates in them.

4-27 *Stain effects.* (From Understanding Wood by R. Bruce Hoadley used with permission of the Taunton Press, Inc. 63 South Main Street, PO Box 5506, Newtown, CT 06470. © 1980 The Taunton Press, Inc. All rights reserved.)

152

Floor finishes

A floor finish is a protective coating designed to give a wood floor a hard, durable surface so it can withstand the rigors of modern living. These products provide a moisture-resistant barrier that will keep the floor looking good for many years to come.

Urethanes are the most commonly used floor finishes on the market today. Moisture-cured, water-based, and oil-modified urethanes are all convertible finishes. A *convertible* finish is cured by a process called *polymer cross-linking*. In this process, the polymers cross-link to form a tight grid as the finish dries. The resulting film is a hard, polymerized, moisture-resistant coating. Moisture-cured urethanes absorb moisture and humidity from the air to accelerate the curing process. They tend to cure more quickly when the atmosphere is damp or humid. These products are highly flammable and contain significant levels of volatile organic compounds (VOCs). They are available only in glossy finishes. And their application should be done by a highly skilled professional.

Oil-modified urethanes are fast-drying, easy-to-apply floor finishes. Commonly referred to as *polyurethane,* these products contain high levels of VOCs and require sanding in between coats. They come in glossy and satin finishes, yet tend to turn amber with age.

Water-based (or waterborne) finishes are currently the most widely used products. Heavily researched and developed over the past decade, waterborne finishes are environmentally responsible because they are nontoxic and low in odor and are typically formulated to comply with most VOC regulations throughout the country. Waterborne finishes are often harder and more abrasion-resistant than some oil-modified finishes. Most waterborne products are nonflammable, nonyellowing, clear finishes that have superior drying times. These finishes provide excellent chemical resistance that protect against water, alcohol, detergents, and a variety of household products. Water-based finishes are generally combinations of urethane and acrylic, along with some kind of catalyst mixed with the finish immediately prior to application. Although they are applied in a similar manner to solvent-based finishes, they typically require more coats to produce the same finish thickness.

Swedish finishes (acid-cured urethanes) contain formaldehyde and are convertible finishes cured by the addition of a hardener. Swedish finishes are likely to bring out the reddish tones in many of the species to which they are applied. They also have a high VOC content, are difficult to apply, and require a meticulously sanded surface. This finish should be applied by a flooring specialist who is familiar with this product and its contents.

Applying the finish

Every manufacturer has specific instructions for applying its products. Read the directions carefully. Pay particular attention to the safety instructions. Since all products require some kind of special handling, the following information is merely general tips regarding the application of floor finishes.

Begin applying the finish in a corner and work toward an area that provides an escape route. Plan your course carefully so that you do not have to cross a wet surface to leave the room. Always work away from the light in the room, so that you can see the wet surface and notice any *laps* (heavy lines) or *holidays* (missed areas). Brush on the finish in long strokes that go *with* the grain of the wood. Overlap each previous row by 4 to 6 in to provide an even finish.

Be ever-vigilant for puddles and debris in the finish. Remove any foreign matter immediately. Continue to apply the finish until the entire floor is covered. When the floor is completely covered with finish, seal the room and let the finish dry. Once the floor has dried, ventilate the area thoroughly.

Many finishes require the floor to be smoothed with an abrasive pad prior to the application of a second coat. This procedure will flatten any raised grain, loosen any debris embedded in the finish, remove any lap lines, and provide a roughed-up surface for the second coat to attach to. Vacuum the floor thoroughly after the sanding is completed.

Follow all manufacturers' instructions for applying the second coat of finish. Spread on the finish just as you applied the first coat. When it has dried completely and you have inspected it for possible irregularities, take any corrective measures needed at this time. Fill any holes or blemishes with an appropriate repair putty, and touch up any areas as necessary. If everything has gone according to plan, you should now be looking at a beautifully finished hardwood floor.

Bleached floors

Wood floors that are a bleached, white color or a pastel shade have gained greater acceptance over the past decade. Light-colored floors blend nicely with contemporary home furnishings and accessories. Although these floors are unique, they do present some potential problems that should be addressed prior to the installation. These are some of the possible complications:

- *Visible cracks exist between flooring strips.* Due to the expansion and contraction of the wood caused by humidity changes, light-colored floors accentuate this problem more than natural or dark-colored wood floors.

- *Bleaching a wood floor is somewhat risky.* It can soften the surface of the wood, create adhesion interference between the wood and the finish, or cause color tone variations.
- *Light-colored floors are high-maintenance floors.* Because soil and wear will be more apparent, regular maintenance is crucial to the appearance of light-colored floors.

By discussing these eventualities with your customers ahead of time, you will create informed buyers. Once those buyers are appraised of the possible hazards of installing this particular product, they can make the decision that is appropriate for them. In addition, they can mentally prepare for how the floor will ultimately look and the level of maintenance that will be required.

Prefinished solid oak flooring

Solid $\frac{3}{4}$-in wood flooring is sold as not only an unfinished product but also a prefinished product. Strip and plank wood flooring can be purchased with factory-applied stains and finishes. They are set in place just as unfinished solid-wood flooring products. Most mills use select grade or better woods that are kiln-dried, because this produces a greater degree of consistency throughout the boards. Precision-stained and coated with several layers of finish (usually urethane), these products are durable and quite attractive.

Prefinished solid-wood flooring is an ideal choice for installations that require minimal dust or noxious odor emissions. Individuals sensitive to these conditions are better suited to a prefinished floor. In addition, in situations where installation time is critical, prefinished floors can be walked on immediately. There is no downtime between the laying of the floor and the finishing of the floor as there is with unfinished flooring product.

The overall cost of the project might be somewhat less with the prefinished version because the extra labor cost for sanding and finishing the floor is eliminated. Yet, the retail price of a prefinished wood product is understandably higher than that of an unfinished floor, so the cost differences are somewhat negated. Therefore, choose the flooring product (prefinished or unfinished) that best suits the particular job requirements.

One of the main problems, however, with prefinished products is that there is a greater likelihood of *overwood*. Moreover, since there will be no opportunity to sand any raised areas as you would on an unfinished floor, greater care must be taken in the preparation and installation stages of prefinished wood floors. Subfloors must be extremely level and flat. Any adjustments to the subfloor have to be done in advance, before you lay any flooring. There will be few, if any, adjustments that can be made later.

Properly installed, prefinished solid oak floors can provide many years of service. If necessary, they can also be sanded and refinished at a later date to bring back the color and shine.

Laminated wood flooring

The floor covering industry, in recent years, has confounded the use of the terms *laminated* and *laminate flooring*. Traditionally, the word *laminated* referred to a product made entirely of real wood materials that was not a solid piece of wood. Recently, however, the term *laminate* is being used to mean melamine-type products that are fused together onto a particleboard-style, or high-density, backing material to form a floor covering that surprisingly (because of photo-imaging) resembles a hardwood floor. (Laminate floors are discussed in Chap. 6.) Because of this confusion, laminated wood floors are now also referred to as *engineered* wood flooring.

A laminated (engineered) wood floor consists of two or more wood veneers that are glued together to form a strong, stable floor (Fig. 4-28). It is constructed in the same ply fashion as a plywood is manufactured, yet it has a top wear surface that is made of a selected, good-quality hardwood (often oak) veneer. The grain directions of all the layers are laid perpendicular to one another to give the floor added strength. Laminated wood products are more dimensionally stable than solid oak floors because they are purposefully *engineered,* if you will, to resist atmospheric moisture and humidity changes by this deliberate cross-grain pattern of the plies. Since wood tends to expand and contract across its grain direction, by laying the grains of the plies perpendicular to one another, as one layer is attempting to expand or contract in one direction, it is being held back by the adjacent ply's trying to expand or contract in the opposite direction. This makes for a wood floor that can even be installed on a concrete slab that is below grade level.

In addition, laminated wood floors can be installed without the large expansion gap around the perimeter of the room required with solid floors. Although it is still necessary to have a gap, it does not have to be an excessive one. (Check the manufacturer's specifications regarding an acceptable expansion space.)

Laminated wood floors can be glued to a concrete subfloor, nailed (using a specially designed nailer) (Fig. 4-29) or glued to a plywood subfloor, or floated over either surface. (The floating installation method is covered in detail under laminate floors because that is the

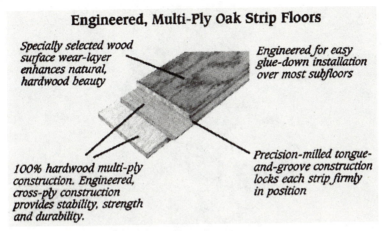

Engineered, Multi-Ply Oak Strip Floors

Specially selected wood surface wear-layer enhances natural, hardwood beauty

Engineered for easy glue-down installation over most subfloors

100% hardwood multi-ply construction. Engineered, cross-ply construction provides stability, strength and durability.

Precision-milled tongue-and-groove construction locks each strip firmly in position

4-28 *A laminated (engineered) wood floor.* (Courtesy of Bruce Hardwood Floors.)

only way laminate floors are installed. See Chap. 6. The majority of laminated floors installed are prefinished. Factory-applied stains and finishes make these floors an ideal choice not only for the professional, but also for the do-it-yourselfer. A homeowner can conceivably transform an uninspiring room to a showcase suite during the course of a weekend. Unrestricted by bulky sanding and finishing equipment, the prefinished wood products have brought the wonders of wood flooring to the average consumer. Since wood increases the value and appearance of any home, it is a logical choice for a remodeling project.

Most laminated wood floors range in thickness from $\frac{3}{8}$ to $\frac{1}{2}$ in. Choosing one of these floors may be necessary when height elevations are a problem. In an area where a $\frac{3}{4}$-in solid strip floor may be unacceptable, such as a low outside-door threshold, the use of a laminated product may be ideal. Therefore, check all areas where the floor will abut before you make a final wood flooring recommendation to your client.

One of the most important features of a laminated wood floor for a contractor is its speed of installation. Often many small jobs can be completed in 1 or 2 days. The ability to lay a great number of square feet, in a shorter time, can only increase your profit margins.

Parquet wood flooring

A welcome relief from the linear look of a strip or plank floor is parquet wood flooring. *Parquet* has come to mean any wood floor that is pieced together as a unit and whose grain is arranged with directional

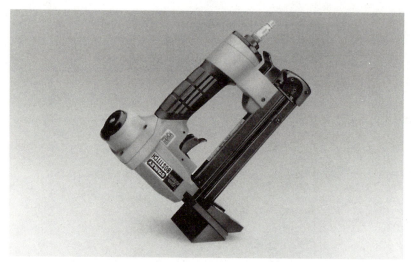

4-29 *Laminated floor nailer.* (Courtesy of Stanley-Bostitch.)

change within the unit. Parquets are typically thought of as the traditional six-finger, alternating-block patterned wood floors. However, there really are a great many more styles from which to choose. Some of the more popular patterns are Monticello, Haddon Hall, Canterbury, and herringbone as well as a host of others (Fig. 4-30). Technological design capabilities make parquet floors more versatile than ever before. Once installed, these floors often look like true works of art.

Although most parquets are $\frac{5}{16}$ to $\frac{1}{2}$ in thick and are meant to be glued down, there are some $\frac{3}{4}$-in solid wood products that can be nailed into a wood subfloor. However, these $\frac{3}{4}$-in products can be installed only on grade or above grade. The thinner versions, however, are the ones that can be installed below grade. Also the directional grain change gives parquet floors great dimensional stability. Consequently, its use below grade is not a problem, providing the slab is dry.

The majority of parquets come factory-stained and -finished. Ease of installation is a big plus. However, there are unfinished parquets available that can be sanded and finished on the job. Make certain the adhesive is thoroughly set up before you use the sanding machine, because the bond with the adhesive can be easily broken if the floor is sanded too quickly.

Wood flooring over a radiant-heated concrete substrate

Some wood floors can be installed over a radiant-heated concrete slab. The preferred method entails gluing laminated plank or parquet

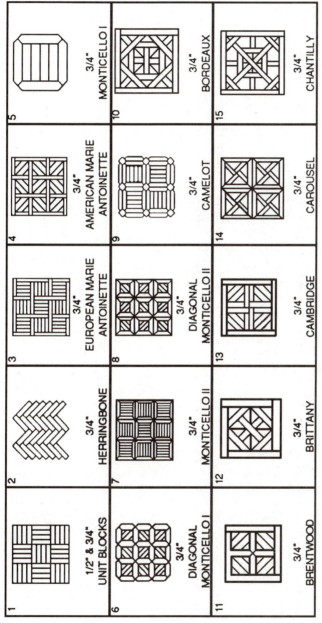

4-30 *Custom parquet flooring chart.* (Courtesy of HG Roane Company.)

1 1/2" & 3/4" UNIT BLOCKS
2 3/4" HERRINGBONE
3 3/4" EUROPEAN MARIE ANTOINETTE
4 3/4" AMERICAN MARIE ANTOINETTE
5 MONTICELLO I
6 3/4" DIAGONAL MONTICELLO I
7 3/4" MONTICELLO II
8 3/4" DIAGONAL MONTICELLO II
9 3/4" CAMELOT
10 3/4" BORDEAUX
11 3/4" BRENTWOOD
12 3/4" BRITTANY
13 3/4" CAMBRIDGE
14 3/4" CAROUSEL
15 3/4" CHANTILLY

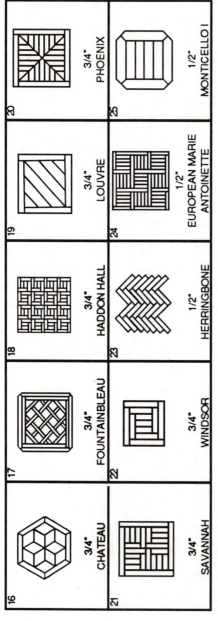

16
3/4"
CHATEAU

17
3/4"
FOUNTAINBLEAU

18
3/4"
HADDON HALL

19
3/4"
LOUVRE

20
3/4"
PHOENIX

21
3/4"
SAVANNAH

22
3/4"
WINDSOR

23
1/2"
HERRINGBONE

24
1/2"
EUROPEAN MARIE
ANTOINETTE

25
1/2"
MONTICELLO I

4-30 *Continued.*

flooring onto the subfloor. In addition it is possible to nail solid-wood flooring into 2-in × 4-in wood screeds that are installed on top of the concrete slab. Still another possibility is to install ¾-in plywood into screeds that are placed 9 to 12 in on center and then to nail into the plywood. However, when you use any type of nail near a radiant-heated floor, take special care not to pierce any of the radiant heat coils embedded in the concrete slab.

When you glue a wood floor to a radiant-heated subfloor, one of the most important considerations is the slab temperature. To eliminate the possibility of adhesive failure, the slab surface temperature should not exceed 85°F. Read all wood product and adhesive company specifications to verify its acceptability with radiant-heated floors. Use of an incorrect adhesive could cause the entire floor to break down.

Disposal of job finish materials

Hardwood floor stains and finishes are a contributing agent to ozone pollution. The U.S. Environmental Protection Agency has identified the volatile organic compounds found in these products. Consequently, proper disposal of these materials is an important part of any flooring installation. Stringent national and state laws have been passed that require users to follow certain guidelines when disposing of these substances. Products that fall into this category are imprinted with this warning: "Dispose of in accordance with federal, state, and local authorities."

Most manufacturers are reformulating the components within their products to comply with current, as well as future, regulations. The increased use of water-based finishes has been a step in the right direction. Oil-modified finishes are much more toxic and harmful to the environment. They should be taken to the appropriate disposal site when they are no longer needed. Remember, these regulations are the law. Fines or other penalties can be imposed if the regulations are not adhered to.

Regional installation considerations

What works in Maine may not work in California. Installation tricks that have been handed down for as many as three generations in Cape Cod might be thought of as lunacy in another part of the country. Since moisture and atmospheric changes greatly affect wood flooring, the presence of snow, frost, or heavy rains will require some specific adjustments.

It is necessary to study your regional market and follow the accepted installation practices best suited for that locale. Trying to incorporate a style from another vicinity could prove disastrous. Listen to a seasoned professional's opinions to understand your local environment.

Care and maintenance of wood floors

In the past decade, wood floors have gained acceptance for installation in areas of the home that were previously unheard of. The installation of wood flooring in kitchens and breakfast rooms seems to be extremely popular today. With the greater likelihood of stains from spills and heavy traffic in these rooms, proper maintenance is critical to preserve the floor's appearance.

Preventive maintenance begins at the point of sale. It is necessary to educate and train all sales personnel about the appropriate products required for any given floor. Improper care instructions can be more harmful than you can imagine. For example, suitable maintenance practices for a urethane-finished floor will be unacceptable for a wax-finished floor. It is always a good idea to provide the essential sundries with every hardwood floor sale. Try to put together a package of products that includes a pertinent maintenance pamphlet, a bottle of cleaner, and a stain touchup kit, if appropriate. Given as a gift, or merely added to the sale, this gesture will not only show your concern for the customers' floors but also will get them started using the right sundries before any damage is done.

Here are some of the basic rules and guidelines for hardwood floors:

- Vacuum, sweep, or dust-mop weekly to remove grit that can cause scratches on the finish.
- Clean up all spills immediately.
- Never intentionally use excessive water on a wood floor.
- If the floor has a wax finish, wax at least once or twice a year. (Heavy-traffic areas may require more frequent waxing.)
- Use walk-off mats at all exterior entrances to remove any grit or dirt that can be tracked in from outside.
- If the floor has a factory-applied finish, use the manufacturers' recommended sundries. (Since they have researched their products, they know what is best to use. Moreover, use of other products may negate any manufacturer's warranties.)

- Use throw rugs in areas that have a high potential for spillage.
- Use soft furniture glides on chairs to prevent scratches and scuffs whenever possible.
- Read all manufacturers' maintenance pamphlets and follow the instructions as closely as possible.
- Repair any damage to the floor as soon as practicable.
- Warn about the danger of wearing spike or stiletto high-heeled shoes. Footwear of this nature can dent or damage a hardwood floor. These shoes can generate dynamic loads that exceed 1000 psi. Shoes of this type, if in a state of ill repair, can be especially ruinous.
- Monitor humidity to maintain recommended levels. When humidity levels are low, use a humidifier to prevent excessive shrinkage in the wood. It is recommended that a humidity level of 45 to 55 percent be maintained. Conversely, in damp conditions use a dehumidifier to keep humidity levels within an acceptable range.

Wood floors are among the easiest flooring surfaces to maintain. They are also one of the most durable. In some medieval castles where the outside stone steps are worn concave from centuries of use, the wood floors inside remain in superb condition. In a modern setting, with the aid of technologically advanced finishes, a properly maintained hardwood floor can last for many, many years.

Acrylic-impregnated wood floors

As discussed earlier in this chapter, wood is a natural product that has pores, much as a sponge does. Acrylic impregnation is a chemical process whereby liquid acrylic is infused into the wood, thus filling the open pores (Fig. 4-31). This procedure bonds the wood and the acrylic together forever. It creates a wood floor that is exceptionally

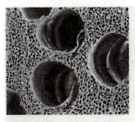

Cross-section of unimpregnated oak hardwood

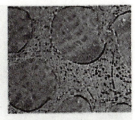

Cross-section of impregnated oak hardwood

4-31 *Acrylic-impregnated wood pores.* (Courtesy of Bruce Hardwood Floors.)

strong and durable. Acrylic-impregnated floors have such a hard wear surface that they can actually be specified for heavy commercial areas. Department stores and shopping centers are ideal candidates for this type of flooring. For residential clients, acrylic-impregnated floors can be used for kitchens, entries, family rooms, or high-traffic areas. These floors are also more scratch-resistant than regular wood floors because of the acrylic infusion.

Custom hardwood flooring

Custom hardwood flooring, also known as *ornamental flooring,* is the product of choice for clients who wants a truly unique floor. A one-of-a-kind beauty can be manufactured by either the traditional hand- and saw-cut method or by the more modern laser-cut inlay technique. Either procedure creates a floor that is both distinctive and decorative. Computer-driven lasers are able to mass-produce intricate wood flooring designs and patterns, thus making these beautiful floors more available to the average consumer. Lasers precut the pattern at the factory. It is then assembled at the factory before it is delivered to the job. After proper acclimatization on site (perhaps for up to 4 days), the floor is installed and finished on-site.

The limit to the design of the pattern is restricted only by the imagination of the designer. The design can be a simple border or an intricate medallion. It can include only two species of wood, or it can include several species. A floor can be comprised entirely of wood products, or it can include what is now referred to as *multimedia flooring.* This is a floor installation that incorporates two or more different raw materials. Some of the products that can be used successfully with wood include inlaid brass, stone, marble, metal, and even semiprecious stones. Since a laser can cut accurately to within $\frac{1}{5000}$ in, it is used to construct elaborate details such as leaves and other radius designs. Also, because they are computer-programmed, they can accurately repeat any established design. It is this programming that gives them their mass appeal because it lowers the cost of production considerably.

Once the design is selected, it is made up into a computer-generated drawing that is then brought to the customer for approval. If the customer accepts the design, the drawing is engineered and *toolpathed* to precise specifications. As with a puzzle, the design is reduced to smaller pieces. If there is more than one species of wood to be used, each variety is cut and separated to ensure proper color placement. The pieces are hand-arranged and held in place by clear face tape. They are brought to the job and installed either as pre-assembled units or as individual, smaller sections. The floor is then finished on the job.

Hand- and saw-cut custom wood floors have a history of quality installations and superb designs. *Kentucky Wood Floors* has long been one of the innovators and leaders in this field. Their award-winning designs have graced not only the floors of many mansions, but also 10 rooms in the White House. According to John P. Stern, president of Kentucky Wood Floors, the use of custom accents and borders is on the rise. Style-conscious consumers are always looking for a way to express their individuality and personal tastes. Reinvesting in the beauty of their own homes is a logical way to exhibit that desire.

Kentucky Wood Floors uses mostly quarter-sawed wood because of its great dimensional stability. When the humidity is high in the summer, it will not expand. Also, when the weather is cold, wood cut in the horizontal plane will not show hairline cracks. Moreover, the straight grain looks better than any other cut of wood and is therefore the preferable choice.

There are over 200 species of wood from which to choose. Although oak is the most widely used species, a few of the many choices include walnut, maple, purpleheart, teak, Brazilian cherry, ash, wenge, and zebrawood. Since all these products vary in porosity and hardness, each accepts a stain (and/or finish) differently. The numerous colors and shades of the various woods also give the designs their unique characteristics.

The patterns and designs are endless. Stock designs like Monticello and Hadden Hall are always a popular choice. Yet custom accents like those shown in Figs. 4-32 to 4-35 are truly stunning. What better way to showcase a room than by installing these imaginative designs right in the floor itself. No rug or other floor could ever have so dramatic an effect. It is one of the most striking approaches to a total design concept in a building that anyone could possibly conceive of for a floor covering.

The patterns are cut into parquet blocks. These blocks range from 8 to 36 in square and can be $\frac{5}{16}$ to 1 in thick. Each block is handmade. The pattern is set into a form that is hammered tightly into place by a skilled artisan. A specially designed latex adhesive is spread across the back of the form to keep everything in place. A vinyl alkyd sealer is spread on prior to the floor's being sold. The flooring is either partially or completely factory-preassembled to facilitate ease of installation. On the other hand, to painstakingly install individual pieces of cut wood flooring on a job would be tedious and very costly.

4-32 *Custom wood floor.* (Courtesy of Kentucky Wood Floors.)

The preassembled forms are usually shipped to the job site unfinished and with square edges. At this point, it is up to the contractor, or owner, to have the floor sanded and finished. If this approach is not acceptable, several options are available to the client upon request: The design can come with beveled edges, a wire-brushed surface, or a hand-distressed texture. In addition, the floor can have a factory-applied penetrating oil sealer, or it can be finished on-site with polyurethane by the manufacturer of the custom wood flooring product.

Coming up with the right combination of design, wood species, and final finish is a long and laborious process. Since these floors are truly custom floors, no two installations are ever alike. Furthermore, whether the materials are laser-cut or saw-cut, the result will be an eye-opening masterpiece. In the end, the visual reward will be worth the effort, and price.

Wood floor possibilities

Wood flooring presents innumerable possibilities of both form and function. There are a variety of species of wood from which to choose, and unlimited designs can be created. The profit potential for the flooring contractor is enormous. Solid-wood flooring, and laminated products as well, each caters to its own unique audience. Homeowners who have concrete floors or who have rooms that are situated below grade can still have beautiful hardwood floors by using laminated wood floors. Conversely, someone who wants a traditional floor may choose solid oak floors. Either floor type increases the value of the home where it is installed. A carefully selected wood floor can truly make a house a home.

Warranties

Warranties for wood flooring come mainly with the factory-prefinished and factory-laminated variety. Since these products are engineered and finished during the manufacturing process, they are more likely

4-33 *Custom wood floor.* (Courtesy of Kentucky Wood Floors.)

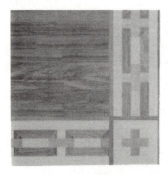

St. Croix

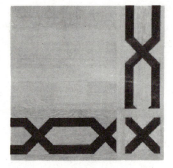

Chelsea

Florentine

4-34 *Custom wood floor.* (Courtesy of Kentucky Wood Floors.)

to carry some good guarantees. Some products will have finish coat warranties, stain warranties, sanding warranties, or overall product warranties. Read all accompanying warranty brochures carefully to verify what features are covered under the express warranty, because each product is unique. No two products and/or manufacturers provide the same coverage. Contact the appropriate manufacturer for any warranty information.

4-35 *Custom wood border.* (Courtesy of Kentucky Wood Floors.)

5

Ceramic tile

Ceramic tile: those two words alone elicit thoughts of beauty, style, and durability. A ceramic tile floor, properly installed and maintained, should last for the lifetime of most structures, because modern technology now provides tile materials that are stronger and more long-lasting than ever before.

Traditionally, ceramic tile was confined to use in bathrooms and kitchens. Today, it is being used more often in areas that were previously covered with carpet or resilient floor covering. For this reason, the remaining years leading into the 21st century will see even further growth in the ceramic tile industry. Homeowners and business owners alike are looking to ceramic tile as a means of covering floors with a relatively permanent product. Even though ceramic tile may be slightly more expensive to purchase and install than, say, carpet or resilient flooring, its longevity is considerably greater. Moreover, ceramic tile has the potential to look better than most other floor coverings for a much longer time.

Ceramic, and stone tile products as well, give the floor covering contractor additional opportunities to increase profits. If you expand into this growing market, one more portion of a building can be added to your floor covering proposal. With the demand for these products growing by leaps and bounds, the innovative patterns, styles, and design motifs make ceramic tiles one of the most versatile floor coverings on the market today. Ceramic tiles are much like custom-designed floors in that the design possibilities are endless. When a truly enduring floor is needed, ceramic tile is the answer.

Tile formations

The use of sun-baked tiles dates back many centuries. Early people saw tiles as a viable alternative to dusty, earthen floors. As tile making progressed and people realized that heating tiles in fire produced

171

much stronger tiles, special ovens called *kilns* were devised to bake the tiles. This process is called *firing*.

Modern tiles are manufactured using any combination of the following raw materials: clay, gypsum, ground shale, vermiculite, talc, and sand. Once the proper mixture has been achieved, the tile is formed into a *bisque*. The term *bisque* is a French word meaning "biscuit," which probably refers to the color ranges of the dried clays. Once the *unfired bisque,* called *greenware,* has been formed, it must be baked in a kiln to improve its hardness. The number of times it will be fired is dependent on the quality of the tile. The temperature for firing tile can be as low as 900°F to as high as 2500°F. The temperature will determine the density and porosity of the tile and glaze. The higher the temperature, the denser the tile and the harder the glaze.

How ceramic tiles are manufactured

There are four basic processes used to manufacture ceramic tile: dust press, extrusion, slush mold, and ram press.

In the *dust-press* method, the tile body is dried to such an extent it is nearly dust before it is formed into shape. The *extrusion* technique forces a much wetter clay body through an opening that is designed to produce a certain shape tile. In the *slush-mold* process, a mixture that is even wetter than that required for extrusion is poured into molds that shape the tile. The *ram-press* system utilizes methods similar to those of the dust-press process but is for larger tiles.

Tile glazes

The vast majority of tiles manufactured today, except perhaps Mexican pavers and quarry tile, have a glaze that is either sprayed or brushed onto the surface. This glaze is applied either prior to firing or after the tile has been fired at least one or more times. The glaze is an opaque or transparent film made up of water, silicates, and pigment. The glaze is designed to add color, texture, and protection to the face of the tile. It can produce a matte, semimatte, or bright glossy finish on the tile.

Tile and water absorption

There are four basic types of tile bisque: nonvitreous, semivitreous, vitreous, and impervious. Each type of bisque absorbs water at a different rate. When a tile is fired, air pockets are formed that determine a bisque's porosity level. That level is established by the temperature in the kiln and the time that the bisque remains in the kiln. In addition, the composition of the bisque itself and its components will be a factor in measuring the water absorption level within the air pockets. Since the glaze that is applied to the surface of the tile absorbs no moisture, it is the body of the tile that is of concern in the firing stage.

The choice of appropriate tile for any given installation will largely depend upon its water absorption level. For example, tiles with a high absorption level are unsuited for installations where water will likely be present, because in this situation a tile that does not absorb much moisture is preferable.

Nonvitreous tile

Nonvitreous tile has the highest rate of water absorption of all the bisque types. It can absorb as much as 7 percent of its weight in water. Consequently, this type of tile is best used indoors where it will come in contact with little or no moisture. If used outside, particularly in cold, damp weather, the tile has a good chance of cracking when the cold air causes the water within the tile to expand as it freezes. Nonvitreous tiles are typically fired at low temperatures for a short time.

Semivitreous tile

Semivitreous tile has a lower absorption rate than nonvitreous tile (from 3 to 7 percent). Even though it is fired at approximately the same temperature as nonvitreous tile, it is left in the kiln for a longer period. It is not recommended for outdoor installations because it still absorbs too much water.

Vitreous tile

Vitreous tile is an ideal choice for almost any type of installation. Being fired at a temperature of about 2200°F for up to 30 h gives this tile its hardness and low water absorption rate. It absorbs only 0.5 to 3 percent of its overall body weight. Also, because of its density, it has high compression strength, which allows it to withstand heavy weights without breaking. Vitreous tiles are well suited for installations as floors.

Impervious tile

Impervious tiles are virtually waterproof. They absorb less than 0.5 percent of their weight in water and are therefore highly bacteria-resistant. This makes them ideal for hospitals and cleanrooms.

Tile backing systems

Almost all tile, when installed, will have spaces between the pieces. Grout is inserted in the gap to form one smooth surface. Plastic tile spacers are used to create a uniform grout joint between each tile (Fig. 5-1). Some tiles are manufactured with self-spacing lugs incorporated into the sides of the tiles. Another way of achieving consistent grout joints is to use tiles manufactured in sheet format. These tiles are attached on the back by thin mesh grids of paper or plastic. In addition, rubber dots are also used to connect tiles (Fig. 5-2).

Sheet-mounted tile (also known as *ceramic mosaic tile*) is generally made up of small tiles held together by a backing to facilitate installation. If an installer were required to lay each individual tile of, say, a 1-in hexagon pattern, it would take far too long. So it's easy to see that sheet-backed tile not only produces consistent grout lines, but also reduces labor time considerably. The backings are left in place during installation to help accomplish these two objectives.

Large, individual tiles, which are not held together by mesh, have a series of deliberately raised areas on the back. Depending upon the individual manufacturer, the raised portions can be in the form of

5-1 *Ceramic tile spacers.* (Courtesy of Doug Messick.)

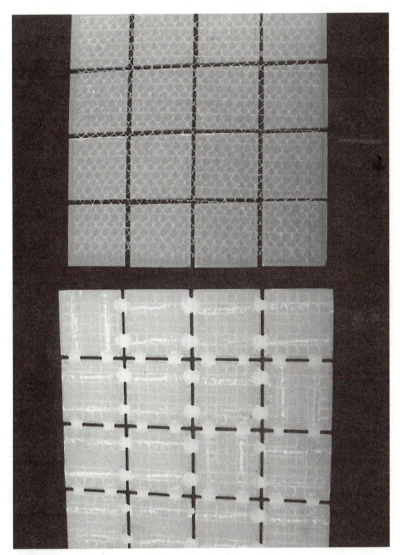

5-2 *Ceramic tile mesh grids.* (Photo by Mahendra Ramawtar.)

dots, ridges, or squares. The raised sections allow for more adhesive to come in contact with the back of the tile, which in turn increases its bonding ability. Also, the raised portions enable the separation of the tiles, when stacked, during the manufacturer's firing process. These elevations provide the opportunity for heat to reach all the surfaces (front, sides, and back of each individual tile), guaranteeing an even firing.

Types of floor tile

The most common types of floor tile are quarry tiles, pavers, glazed tiles, and patio tiles. These tiles all come in various shapes, styles, and colors. They are either glazed or unglazed. Floor tiles are extremely durable due to the raw material content of the tile itself and the firing process that the tiles undergo.

Quarry tiles

Quarry tiles are semivitreous to vitreous clay tiles that are often left unglazed. Available in a variety of shapes (from 4- to 12-in squares, hexagons, and rectangles), quarry tiles can be used for both indoor and outdoor installations. If desired, quarry tiles can be finished with a sealer. A top-coat sealer made especially for quarry tile should be used and applied after the tile is installed, but before the grout is spread into place.

Pavers

Paver tiles are typically made of clay, shale, or porcelain. They are at least $\frac{1}{2}$ in thick and come glazed or unglazed. Paver tiles can be either machine-made or handmade. These tiles are usually extremely durable and water-resistant.

Many of the unglazed handmade pavers are products of the Mediterranean region or of Mexico. Some of these tiles, however, are poorly made due to inferior firing methods and are therefore susceptible to cracking and breakage. The surfaces and backings of handmade tiles are generally not flat. This lack of uniformity requires the mastic to be buttered onto the backing of the tile to ensure proper bonding into the setting bed. Like quarry tile, unglazed pavers can be sealed prior to grouting to protect the tile from premature wear. The sealer also protects the tile surface and helps facilitate the grouting process by making the top of the tile slicker and less likely to absorb moisture from the grout.

Glazed tiles

Glazed floor tiles, either foreign or domestic, are some of the most widely used tiles today. Found in various sizes from 4 in × 4 in to 12 in × 12 in or greater, the most commonly used size is 8 in × 8 in. These tiles can be obtained with any of the available glaze finishes: matte, semimatte, or glossy.

Patio tiles

Patio tiles are usually thicker than most other tiles, up to 1 in, and are generally nonvitreous. They also have more irregular shapes and sizes than other tile products. Because they are nonvitreous they may have a tendency to crack if installed outside in a cold climate.

Tile grading system

Tiles sold in the United States are graded by a system established by the American National Standards Institute (ANSI). Even if tiles are made in a foreign country, they are still graded by the same criteria since they are sold in the United States. The system classifies tile quality as standard grade, second grade, and decorative thin wall tile.

Standard-grade tiles are tiles that meet all the minimum requirements set forth by ANSI. Second-grade tiles are as structurally sound as standard-grade tiles except there may be slight irregularities in the finish or the sizing of the tile may be imprecise. Decorative wall tiles are those tiles that are structurally unsound in both the bisque and the glaze. They are considered unsuitable for use in a functional capacity. Their use should be limited to application on walls.

Tiles that cannot be classified or used in one of the above-named categories are known as *culls*. Culls are so inferior that they are not usable; they are therefore crushed and utilized as roofing chips, or they are exported. It is unlikely that you would be involved in a project to install a floor using culls.

Choosing the correct tile for an area is the first step toward a good installation. The cost difference from one grade to the next may not be all that great, but the proper tile choice may make all the difference in the world when it comes to durability. Therefore, take the time to analyze all the circumstances surrounding a project so you can guide the consumer to the right product.

Floor preparation

As with any floor installation, the surface to be covered with tile has to be properly prepared. It must be free of all foreign substances such as grease, oil, drywall mud, and paint. It has to be clean, dry, and smooth. The floor itself should not be out of level by more than $\frac{1}{2}$ in in a 10-ft span. If the floor is not level, steps must be taken to provide an acceptable surface. Out-of-level floors may require floating a mortar bed, applying self-leveling compounds, or shoring up the subfloor in some fashion. Regardless of the method chosen, the floor has to be structurally sound and in good condition to receive the new materials.

If the subfloor to be tiled over is a concrete slab, it should have a steel trowel and fine broom finish with no curing compounds used, in order for it to be an acceptable surface. For a wood subfloor, if plywood is to be used, typically it must be $\frac{5}{8}$ in thick or 1 in for nominal boards if they are installed on joists that are 16 in on center. If a *cement backer board unit* (CBU) is to be used, it can be installed over a plywood subfloor. CBUs are mesh- or fiber-reinforced cement sheets that are used as a sturdy subfloor onto which tiles may be applied.

If you make sure that the surface is the correct one, the installation will start off properly. The condition underneath any floor is the key to successful application of the flooring. Inspect all floors prior to the installation of any ceramic tiles so that any subfloor problems can be rectified beforehand.

Setting materials for floor tiles

There are a number of setting materials available on the market that will be suitable for any type of tile installation. The Tile Council of America lists these most commonly used products:

- *Portland cement mortar.* This mixture consists of portland cement, sand, and other additives that are often factory-mixed in bags. Dry-set mortars need only be combined with water, or a liquid polymer, to complete the mixture.
- *Latex–portland cement mortar.* This compound consists of portland cement, sand, and an added latex which increases its strength and bonding ability.
- *Epoxy mortar.* Mortar products in this category use epoxy resins and hardeners to provide a rigid bond between the tile and the substrate. Subfloor surfaces that are difficult to adhere to are best served by using an epoxy mortar setting system.
- *Modified epoxy emulsion mortars.* This system uses a mixture of portland cement and silica sand with emulsified epoxy resins and hardeners.
- *Furan resin mortar.* This setting material employs furan resins and furan hardeners. It is a thin-set method that is best suited for installations where chemical resistance is of the utmost importance.
- *Epoxy adhesive.* These adhesives are also used for thin-set installations and consist of epoxy resins and epoxy hardeners.

- _Organic adhesives_. This mixture is factory-prepared using organic materials. It is designed for interior use only and comes in ready-to-use containers. It does not require the addition of water or other liquid, powder, or latex. It is the evaporation process that causes the adhesive to cure and set.

The type of setting material that is best suited for any particular job will depend upon the substrate involved. A certain product that works well over one kind of subfloor will be virtually ineffective over another. It is also important to consider the amount of traffic a particular area will receive and whether it will be exposed to any water. Rooms such as bathrooms and laundry rooms that run the risk of water spillage and dampness should be totally waterproofed. This precaution will protect the owners' original investment by increasing the life of the floor tile installation. Consequently, adhesive and setting-bed compatibility are crucial.

Harmony between the substrate and the adhesive is the single most important factor in any tile installation. For example, when CBUs are used, tiles can be stuck to it using dry-set, latex–portland cement, or modified epoxy mortars. On the other hand, an installation over sheet vinyl flooring requires an epoxy thin-set adhesive. (The term _thin set_ refers to mortar-based adhesives which are cement-based powders that are combined with a liquid prior to application. There are water-mixed, latex, acrylic, or epoxy thin sets that provide superior bonds and flexibility compared to most organic adhesives.)

Prior to purchasing any adhesive, you should always consult the manufacturer or retailer to make sure it is the appropriate kind of setting material for the particular substrate. Most professional floor covering people are more than willing to answer questions about products they produce or provide. Read all instructions and warnings on the package before you proceed with any tile installation.

Grouts

Grouts, like setting materials, come in a variety of forms. The type that is the most appropriate for a specific installation will depend upon the style of tile and the kind of installation (wet or dry). Grout is essentially another form of setting bed. It is typically a premade powdered mixture that is combined with a liquid to form a thick paste. This paste is then forced into the spaces between the tiles to give the entire floor greater stability and support.

The basic ingredient of most grouts is portland cement. Portland cement is then altered to adjust to the quality requirements of specific grouts. If a particular grout has to provide a certain hardness or mildew resistance, modifications can be made, and additives can be included, to meet those requirements. On the other hand, a number of grouts are not cement-based, and they have their own niche in the tile industry because they offer features that cement-based grouts cannot. The types of grouts generally used are as follows:

- *Commercial portland cement grout.* The type used for floors is typically gray and is a blend of portland cement and various other ingredients which form a dense, water-resistant grout.
- *Sand–portland cement grout.* This product is mixed at the job site to varying specifications based on the width of the space between the tiles. A 1:1 mix of 1 part portland cement to 1 part fine-grade clean sand can be used on tiles with a $\frac{1}{8}$ in gap. The ratio can go up to 1:3 for a $\frac{1}{2}$-in gap.
- *Dry-set grout.* Portland cement and special additives are combined to produce a grout with excellent water retentivity.
- *Latex–portland cement grout.* This is a portland cement–based grout with certain latex additives included. This grout is used for most residential and many commercial installations.
- *Epoxy grout.* These grouts use epoxy resins and hardeners and are chemical-resistant.
- *Furan resin grout.* Grouts in this category contain furan resins and hardeners and are highly chemical- and temperature-resistant.
- *Silicone rubber grout.* This grout is an engineered elastomeric system that uses silicone rubber. Care should be taken and the manufacturers contacted before you use this kind of grout in a food preparation area. It is highly stain-, moisture-, mildew-, and crack-resistant.

Installation of ceramic tile

The following guidelines are merely basic steps for setting ceramic tile floors. Note that many variables must be taken into consideration regarding any specific tile installations. Also, there are a number of subtle nuances, and habits, that an installer picks up over the years that aid in the installation process. The aim of this section is to provide fundamental information to help familiarize you with installation procedures.

Tools

The professional's tool box will vary greatly from the do-it-your-selfer's tool box. Since the professional will encounter a wide range of installation problems, and since those same problems or situations may arise more than once, they always purchase the proper tools. The do-it-yourselfer, on the other hand, will rarely encounter these circumstances on a regular basis, so it is more appropriate for them to rent many of the tools from a local rental agency or tile shop where the tiles are purchased. These are a few of the many tools necessary to install almost any ceramic tile:

- Floats
- Measuring tape
- Tile cutters and nippers
- Trowels—flat, margin, and notched
- Straightedges
- Carpenter's level
- Rulers
- Chalk line
- Marking pencil
- Wooden block and mallet
- Mixing pail
- Sponges
- Carborundum stone
- Rubber gloves
- Carbine scriber
- Carpenter's square
- Tile spacers
- Safety equipment

(See Fig. 5-3 for a condensed illustration.)

If a wet saw is needed for a particular installation, it can be rented or purchased. A wet saw is needed for hard-to-cut tiles or tiles less than $\frac{1}{2}$ in wide. Stone tiles, hard-bodied tiles, and soft-bodied tiles can all be cut on a wet saw. A *wet saw* is a table saw that has a diamond-blade cutting wheel. It must be used with water that is shot through jets which are positioned to hit the blade just slightly above the point where the tile is cut. Most wet saws have a sliding table onto which the tile is mounted. The tile is then pushed through a cutting blade that is in a fixed position. The wet saw makes smooth, straight cuts.

Some of the more important safety equipment includes safety goggles, a charcoal-filtered mask, and hearing protectors. Thinking *safety first* is the first rule to follow in any tile installation. Protect yourself and those around you by being prepared and alert at all times.

5-3 *Ceramic tile installation tools.* (Courtesy of TEC Incorporated.)

Tile layout

Laying out a ceramic tile installation is the same as laying out a resilient tile installation (see Chap. 2). Just as with resilient tile, it's a good practice to "dry lay" the tile before spreading any adhesive, to get an idea of how the tile will look when installed. The most important rule is to use as many full tiles as possible throughout the installation. If any tiles need to be cut, they should be no less than one-half of a tile. This is accomplished by dividing the room into quarters by using two intersecting chalk lines and adjusting those lines either vertically or horizontally to accomplish that goal (Fig. 5-4).

When you lay ceramic tile, however, you must allow room for the grout when doing the dry layout. Therefore, use the tile spacers to achieve the correct distances between the tiles.

By doing a dry layout, you get a preview of what the job will ultimately look like. This step eliminates the possibility of any unpleasant surprises later on. It also saves you from having to re-lay any tiles after they have already been placed into the setting material.

Installing ceramic tile

With the room properly divided into quarters by chalk lines, begin spreading the adhesive or mortar. Start from the center point, and spread only 3-ft² areas at a time. Also, concentrate on completing only

one-quarter of the room at a time. Make sure to leave yourself an escape route, so work toward a door to eliminate the possibility of
"painting yourself into a corner."

Dump an amount of setting material onto the floor that you think
you can reasonably work, and begin spreading it with the flat edge of
a trowel held at a 30° angle. Work from the center point of the room
outward into the section you will be installing. Then, with the
notched edge of the trowel, begin *combing* out the setting material
into long ridges (Fig. 5-5). These ridges will provide a better bond between the tile and the adhesive. (Large button-backed tiles may require additional *buttering* of adhesive to the back of the tile to ensure
a proper bond.) When you comb the adhesive, hold the trowel at an
angle anywhere from 45° to 70°.

It is extremely important to use the appropriately notched trowel
for the tile installed. If a large-notched trowel is used for a small tile,
there will be too much adhesive for the tile and it will seep up into
the grout joints. Conversely, a small-notched trowel will not be sufficient to fully transfer enough adhesive to a large lug-backed tile.

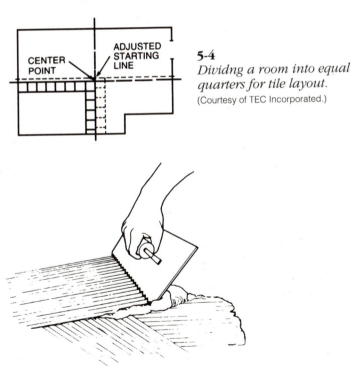

5-4
*Dividng a room into equal
quarters for tile layout.*
(Courtesy of TEC Incorporated.)

5-5 *Combing ceramic tile-setting
material.* (Courtesy of TEC Incorporated.)

Once the adhesive is properly combed into position, begin setting tiles by placing the first tile at the center point of the two intersecting lines. It is imperative to set the tiles rapidly in order to achieve the strongest bond between the tile and the adhesive. Lightly twist the tile as it is being placed in position. Try *not* to slide the tiles into place, because that causes the adhesive to rise up into the grout joints. If adhesive does happen to get into the grout joint, clean it out as quickly as possible. Adhesive left to harden in a grout joint can be extremely difficult to remove. Excess adhesive is relatively easy to clear away if it is removed while still soft. If adhesive is left to harden in a grout joint and cannot be removed, it may cause the grout to crack at that location because the depth of the grout will be too thin.

Continue laying the tiles in one section of the room at a time. After a sufficient number of tiles have been placed in position, use a rubber mallet and beating block to gently beat the tile into the adhesive (Fig. 5-6). This will ensure that sufficient adhesive is transferred to the back of the tile.

If the laying of the tiles goes more slowly than anticipated, the adhesive may begin to *skin over.* (An adhesive skins over when it has been allowed to sit for too long and a film has begun to develop on its surface.) If tile is placed into a skinned-over adhesive, a weak bond will develop and the tile may ultimately lift up. Therefore, if this situation arises, remove the skinned-over adhesive and reapply fresh adhesive.

After the first section (or quarter) has been laid, install the second, third, and fourth sections. Make sure that all the lines are straight, as well as parallel and perpendicular. There is nothing worse than seeing a completed tile installation with crooked or wavy grout

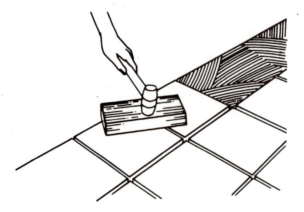

5-6 *Beating block and hammer.* (Courtesy of TEC Incorporated.)

lines. It looks unprofessional and can make even the most expensive tiles look awful.

Once all the tiles have been laid, allow the setting material time to set up before grout is applied to the joints. Read the manufacturer's instructions carefully to find out how long that particular adhesive takes to set up. If grout is applied over adhesive that is still wet or is not fully cured, the grout could become discolored due to excess moisture in the adhesive. If some portions of the room have dry adhesive and others have wet adhesive the grout could look spotty. So wait the appropriate time period before you begin grouting.

Applying grout

Grouting a ceramic tile floor is often as important as the installation of the tiles themselves. A poorly grouted floor can ruin the appearance of a good tile installation. Choosing the right grout to go with the tile will enhance the look of the job.

Grouts come in a multitude of colors, and they can make an otherwise ordinary tile look exceptional. Plenty of commercially premixed grouts are designed to complement almost any tile. It is advisable to wait at least a day or two (depending upon the moisture, humidity, and atmospheric conditions) before attempting to grout the floor, so the adhesive can cure. Once you are certain that it is time to begin grouting, remove any tile spacers in the grout lines. Mix the grout according to the manufacturer's directions and then allow it to set up for about 10 minutes (min). Restir the grout before you use it, to loosen up the consistency a bit.

The grout should be dumped onto the floor in a heap that will cover a small area of approximately 15 to 20 ft². Work in small sections, and observe how quickly the grout sets up. If it sets up quickly, you will have to clean it much sooner—to get the tiles clean before the grout begins to harden. Do not try to grout the entire area at once, although sometimes it is possible to grout large areas if the grout is setting up slowly.

Use a rubber float to spread the grout, and hold it at a 30° angle (Fig. 5-7) to the surface of the tile. When you apply the grout, work diagonally across the grout lines and thoroughly pack it into the joints. To remove the excess grout, hold the rubber float at a 90° angle, and again work diagonally across the tile. Working diagonally across the grout lines reduces the probability of having the grout pulled out of the joints. If this does happen, simply refill the joint and let it dry.

5-7 *Rubber float and grout.* (Courtesy of TEC Incorporated.)

Cleaning the grout

After the excess grout has been removed and the grout has had sufficient time to cure, final cleanup may begin. Most grouts can be
cleaned with water and a sponge. Soak the sponge in a bucket of
clean water, and then remove as much excess water as possible. As
in the process of applying the grout, begin by working in a diagonal
direction to the grout lines. Several passes may be necessary to remove the excess grout and the resulting haze, which otherwise will
remain on the surface of the tiles.

Since grout particles accumulate in the pores of the sponge, constant rinsing of the sponge is essential. When one side of the sponge
begins to become soiled, flip it over and use the other side. Then
rinse the sponge thoroughly, and begin the process again. When you
have made sufficient passes to remove the grout and clean the surface of the tile, let it sit for a while to allow the grout to harden. After
this time, wipe the tile with a clean, soft cloth (Fig. 5-8).

It is important to remember that during the cleanup process, the
shape of the grout is being formed. During this stage, it may be necessary to work the sponge parallel to the grout lines to create a uniform grout level. Whether the grout line will be flat on top or be
slightly concave is not critical. What is important is that all the grout
line heights, and the overall appearance of the joints, remain consistent throughout the installation.

Ceramic floor tile maintenance

Once the grout and adhesive have thoroughly cured, you may want to apply a supplemental sealer or impregnator to the tile, to the grout, or to both, to protect against staining. Although acrylic or latex additives or polymer-modified grouts decrease the likelihood of staining, these after-installation sealers are extremely effective.

A regular maintenance program is the best way to preserve the fresh, new look of any tile installation. Since ceramic tile is one of the easiest floor coverings to maintain, often a mere sweeping or vacuuming is all that is necessary. For normal cleaning or more stubborn stains or spills, use a household detergent-and-water solution to clean the floor. The best method is to use two buckets of water; one containing soap and water and the other containing clean water. Work a small section at a time, perhaps 6 ft × 6 ft. After the section has been washed, rinse it quickly to remove any soap residue. Soap residues tend to cloud the surface of the tile and cause the glaze to lose its luster. If this does happen, rewash the floor, rinse, and wipe dry with a soft cloth.

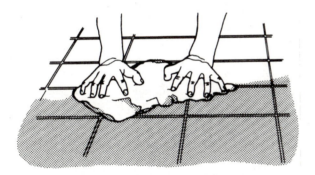

5-8 *Cleaning grout.* (Courtesy of TEC Incorporated.)

6

State-of-the-art floor coverings

As technology has advanced and consumer demand has increased for newer kinds of floor coverings, the flooring industry has responded with innovative products. Among these innovations, two products stand out: laminate floors and the hook-and-loop carpet system.

The enthusiasm generated by laminate floors in the past few years is unprecedented in the flooring industry. Consumer acceptance of these products has been enthusiastic because laminate floors offer so many benefits that make them hard to resist. Ease of installation, exceptional durability, and minimal upkeep have brought laminate floors to the forefront of today's flooring market. When laminate floors were introduced, there were only a few manufacturers, and distribution was limited. Now, as with any successful product, more companies are entering the field. The carpet industry, on the other hand, has recently developed a new installation method. The hook-and-loop carpet system is a product marketed by the 3M Company called *TacFast Systems*. This process requires no adhesives, no stretching, no kicking-in of the carpet, and no hot-melt seam tape. Also, it releases no harmful VOCs into the air. The TacFast carpet system provides all the benefits of a broadloom carpet installation, without all the fuss and mess of the traditional stretch-in technique.

The above two products offer consumers new options to conventional floor coverings. They both offer greater design capabilities along with easy installation procedures. A look at these products will give us a glimpse at the future of the floor covering industry.

Laminate floors

According to its own literature, the company that invented laminate flooring is Perstop Company of Sweden. Their product is called Pergo

Original. A laminate floor is the result of a process of combining two or more substances under high temperature and pressure to create an extremely durable product. The wear surface is made up of melamine resin. These resins are impregnated onto paper to form the top surface layer of a laminate floor. This surface, which is highly wear-resistant as well as decorative, is pressed onto a moisture-resistant chipboard. These two layers are then secured to a bottom layer of laminate for structural stability (Fig. 6-1). Through precise engineering, the three layers of material are formed into tongue-and-groove planks that are 47 in long, 8 in wide, and $\frac{1}{4}$ in thick.

The wear surface is made even stronger by the introduction of hard particles that reinforce its structure. The reinforcing particles give the laminate floor greater wear resistance than that of many of the other top flooring materials on the market today. Even the high heels of women's shoes, which exert a tremendous amount of pressure per square inch on any floor product, will not dent a laminate floor. Household furniture will not permanently dent the surface either.

Melamine resins are so highly heat-resistant that cigarettes or cigars left on the surface to smolder will not mar the finish. In addition, the finish is extremely stain-resistant; nail polish, ink from felt-tip pens, or tracked-in asphalt can be readily cleaned up with acetone or mineral spirits. A laminate floor never needs waxing. Simple vacuuming, the use of a special spray cleaner, or damp mopping is all that is needed to maintain its wonderful appearance.

Most of the laminate floor products come in a wood-look design. However, as other companies enter the market, the design possibilities expand. There are also marble designs, solid colors, and floral patterns in the Pergo line, while the Bruce Flooring Company has added feature squares ($7\frac{5}{8}$ in $\times$ $7\frac{5}{8}$ in) to its new line of laminate floors. These feature squares come in solid colors as well as a granite

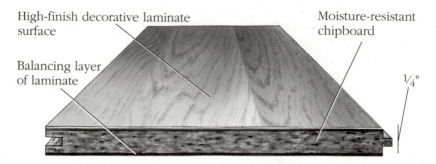

6-1 *Laminate floor crosssection.* (Courtesy of Perstop Company of Sweden.)

simulation. The use of the feature squares can help dramatize the appearance of an otherwise undistinguished-looking floor.

As laminate floors have increased their market share in the flooring industry, there have also been developments in structural materials and surface finishes. The introduction of high-density fiberboard (HDF), instead of chipboard, amplifies the tensile strength, hardness, and stability of the new laminate products. An HDF core has superior impact resistance and water resistance. Also, highly sophisticated photographic equipment is used to produce images for patterns that are virtually lifelike. The definition of pattern in the wood grain is truly remarkable. It is often very difficult to tell the difference between a laminate floor and the real product it emulates.

Installing laminate floors

Laminate floors are installed by using the floating floor method. In this installation procedure, the flooring material is not adhered directly to the substrate. These products are glued together with an adhesive that is applied along the tongue and groove of the *board* only. Most laminates can be installed on-grade, above-grade, and even below-grade. They can be installed over almost any surface of existing floor covering providing the substrate is smooth, dry, clean, and structurally sound. Laminate floors can be installed over concrete, wood, sheet vinyl, or even ceramic tile. Carpeting, however, is the only floor covering which is not recommended under the installation of a laminate floor. It is always best to remove any wall-to-wall carpet before you proceed with a laminate floor installation.

Whenever a laminate floor is to be installed over a concrete floor that is either on-grade or below-grade, a moisture barrier of polyethylene is required. The thickness of the polyethylene film should be between 6 and 8 millimeters (mm). (Each manufacturer has specific requirements for film thickness, so read all necessary instructions first.) If the floor is to be installed on a raised-foundation wood subfloor, the polyethylene film need not be used. What is necessary on wood floors, however, is to ensure that the moisture content of the subfloor is between 6 and 12 percent. Use a moisture reader to measure the moisture levels of the wood subfloor before installation.

One of the most important requirements in a laminate floor installation is that the floor be smooth and flat. It is essential that the subfloor vary no more than $\frac{3}{16}$ in in a 10-ft span. If the substrate does not conform to these specifications, proper corrections must be made. Wherever depressions exist, they must be leveled by using a latex leveling compound. If there are any raised areas, they should be sanded down or repaired accordingly.

As with any tile or hardwood installation, plan and measure the room so that the boards lay out in such a fashion that no less than one-half of a board is left at any wall. If this means adjusting the size of the beginning board, then do so from the start so that the final board is the correct width.

Once all necessary preparation work has been completed, a layer of $\frac{1}{16}$-in foam is rolled out to cover the floor. This foam provides the base over which the boards will float. Working with a three-row span of boards at a time, lay the foam as needed. Since you will be working across it and on it, lay only the amount needed.

The first three rows of any laminated floor installation are the most critical. They must be laid straight and must fit together snugly. After using the appropriate spacers to provide a $\frac{1}{2}$-in expansion gap between the wall and flooring, dry-lay the boards to check the fit. Cut and position the first three rows of board before you apply any adhesive (Fig. 6-2). If any adjustments are needed to be made, it's best to make them before any glue is applied.

Once the beginning rows have been properly laid out, the glue may be applied *only* to the tongue-and-groove portion of the laminate boards. Since this is a floating installation system, no glue is ever applied to the underside of the boards. (Be aware that all laminate floor manufacturers have particular specifications regarding the adhesive application procedure. Some companies recommend inserting the adhesive into the grooved portion of the board's edge while others specify that the glue is to be applied to the tongue area. Read the appropriate set of instructions before you apply any adhesive.)

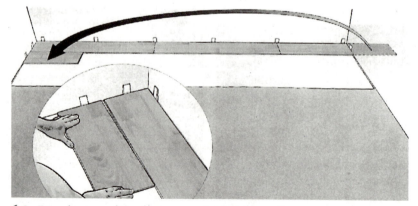

6-2 *Dry–laying of the first rows of a laminate floor.* (Courtesy of Perstop Company of Sweden.)

6-3
Clamping boards. (Courtesy of
Perstop Company of Sweden.)

The use of specially designed clamps will aid in firmly securing the
first three rows of boards (Fig. 6-3). Once the glue has been applied
and the clamps have been set, allow from 30 min to 1 h before you resume the installation of any other planks. Since the setting of the first
three rows is so vital to the success of the rest of the job, you don't
want these boards to move and slide as you are installing additional
rows. Use this downtime to remeasure and cut additional rows of
flooring. When you lay out the floor, make sure that the joints are staggered, ashlar fashion, for the best appearance. Never have an end joint
next to another end joint. That creates a weak area in the floor and
makes the seams look too linear. For strength and proper appearance,
the end joints should be 16 to 25 in from any adjoining end joint.

After the clamps are removed from the first rows, use a tapping
block and mallet to gently tap the remaining boards into place
(Fig. 6-4). Do not hammer directly onto the board itself because that
will cause the plank to become chipped and out of alignment.

One of the most important things to remember when installing a
laminate floor is to clean up the excess adhesive from the surface as
quickly as practical. Allow the glue enough time to begin the bonding process, but not so much time that it begins to harden. Wipe up
the excess glue with a clean, damp rag. If the entire floor is laid before the excess glue is wiped up, the glue may be difficult to remove
without damaging the surface. Abrasive products, such as steel wool
or sandpaper, should not be used to remove anything from the surface of the planks. When the last row of planks is to be installed, remeasure and cut it as you would in any other installation. This plank
will have to be cut lengthwise to fit into place. A crowbar or specially
designed pull bar is needed to secure the last row into the previous
row (Fig. 6-5). Leave enough of a gap between the plank and the

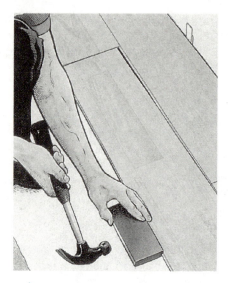

6-4
Tapping block and mallet.
(Courtesy of Perstop Company of Sweden.)

6-5
Pull bar. (Courtesy of Perstop
Company of Sweden.)

wall to insert the spacers. Leave the spacers in place until the floor is ready to be used.

Allow the floor to sit, unused, for 8 to 12 h. No one should walk on the floor until it has had sufficient time to cure. If there is any excess glue remaining on the floor, it can be removed later with acetone.

Remove the spacers and install the wall moldings to complete the job.

A floor for the future

Laminate floors—or *high-performance floors,* as they are becoming known—are truly state-of-the-art floors. These floors are in great demand because they are up to 20 times more durable than a traditional sheet vinyl floor. In addition, the warranties offered with these products far exceed those of most other floor coverings.

High-performance floors are projected to continue increasing their market share well into the next century. Until another technological breakthrough brings forth a product that outperforms the current crop of materials, laminate floors sales will continue to gain momentum. Starting from a mere sliver of sales in 1995, high-performance floors are projected to capture 5 to 10 percent of the floor covering market by the year 2001. It is seen as an emerging growth industry that will benefit both those who are selling the products and those who are installing them. Financial opportunities exist for anyone who is willing to evolve with this new segment of the floor covering field. Because laminate floors are used more often in kitchens than in any other room of the house, buyers are taking away more purchases from the standard resilient floor segment than from carpeting.

Warranties on high-performance floors

Among the best features of the new laminate floors is their unprecedented warranties. Depending upon the manufacturer or particular product, the warranties cover from 10 to 15 years. The warranty typically states that for the period covered by the written promise, the floor will not stain, fade, or wear through the surface of the boards. The warranty also covers any manufacturer's defects related to the actual construction of the product. If any of these problems become evident, the manufacturer will either repair or replace the floor, at the manufacturer's option.

As with any warranty, however, it is essential to know what is not covered. Warranties do not cover surface scratches, chipping, or any surface damage caused by negligence or improper maintenance. It is a common misconception of consumers that laminate products are impervious to damage. When they see that the surface cannot be burned with a cigarette or stained by a felt-tip marker, they invariably think that the surface will not scratch either. This is not true. If the buyer does not use the proper chair leg protectors or drags objects across the surface, the floor can become damaged. Also if the consumer uses cleaning products other than those specified by the manufacturer, the warranty may become void. Consequently, it is the responsibility of the person selling the floor to inform the buyer of these restrictions. By making consumers aware of the exclusions, it will help them protect the floor from the beginning. An informed customer is a grateful customer, so don't be afraid to tell the customer what is not covered by the warranty. A warranty is good only when all the guidelines set forth by the manufacturer in the written contract have been complied with by the consumer.

Hook-and-loop carpet system

Hook-and-loop technology is being introduced to the carpet industry as a new anchoring system for broadloom carpet. Its simplicity and environmental benefits are revolutionary. This installation method is a completely integrated system involving specially manufactured carpet, along with unique installation products.

Tac Fast Systems International of Switzerland developed a carpet backing system that, when combined with a hook tape designed by 3M Company, created a mechanical bond between the two products that made carpet installation extremely easy. A layer of loop fabric is applied to the back of the carpet by the carpet manufacturer. This backing can be bonded to a carpet that has an attached cushion, or it can be applied to a carpet that does not have an attached cushion. During the installation process, this loop fabric is pressed onto a hook tape that has been applied to the subfloor with pressure-sensitive adhesive (Fig. 6-6). The knitted loop fabric (or *scrim*) is substituted for the secondary backing of broadloom carpet (Fig. 6-7). This backing

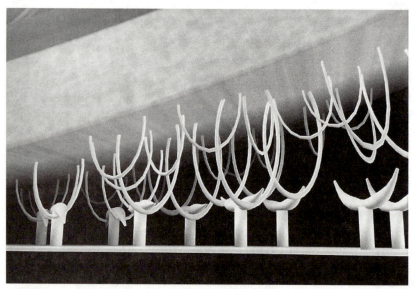

6-6 *Blowup of pressure-sensitive hook tape.* (The TacFast carpet system is marketed exclusively by 3M.)

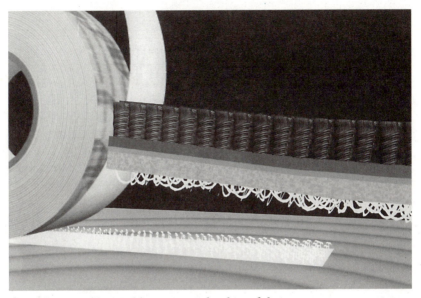

6-7 *Blowup of knitted loop carpet-backing fabric.* (The TacFast carpet system is marketed exclusively by 3M.)

allows for cutting the carpet fibers without fraying. Therefore, custom inlays and borders are much easier to do. Power stretchers, adhesives, tack-strip, and hot-melt seam tape are not required for this type of installation, as they are with conventional carpets and installations.

Installing hook-and-loop carpeting

To install hook-and-loop carpeting, you must purchase the carpeting as well as the hook-and-loop tape. This tape has a fuzzy-fabric, loop-like top surface and a pressure-sensitive adhesive bottom surface. (The tape has a protective covering applied to its top to prevent it from engaging the loops prematurely.) Measure all rooms accurately, and mark the locations of all seams. Surface areas should be clear and clean. A 2-in band of hook tape is placed around the perimeter of the room's floor. Once the perimeter tape is pressed into place, dry lay the carpet to verify exact seam location. With the two carpets butted next to each other, insert a pencil into the seam and mark a line along the entire length of the seam. This establishes a precise seam line. Fold back the two carpet sections. Then a 4-in band of hook tape is applied to the subfloor wherever a seam will be placed. The 4-in-wide hook tape should be centered directly on the pencil line so that there is 2-in of tape on either side of the pencil mark. This will provide the strongest bond to secure the carpet to the tape. Roll the tape into place with a 75-lb roller or a hand roller.

Position the carpet on the tape, making sure that the seams are properly trimmed and that any patterns in the carpet have been properly aligned. One of the best features of this system is that if the carpet needs repositioning adjustments, they can easily be done by simply lifting the carpet off the hooks and making the adjustment. If necessary, a knee-kicker can be used to adjust the seams correctly. When all the seams are fitted accurately together, begin removing the protective covering on the tape at the midpoint of the seam. Slowly lift about 6-in sections of the covering, and press the carpet onto the hooks. Continue to remove the covering and press the carpet onto the tape until you come to the end of the seam in that direction. Then go back to the midpoint and, following the procedure just used, complete the seam going in the opposite direction. When all the seams are completed, roll over the top of them with the 75-lb roller or a hand roller to thoroughly secure the carpet to the tape. Trim all wall edges with a wall trimmer before you remove the tape covering the perimeter of the room. Then gently lift the covering, a few inches at a time, and secure the tape and the carpet along the walls. Finish the job by installing any necessary door moldings to hide the raw edge of the carpet.

Environmental considerations of hook-and-loop installations

As consumers become more and more concerned about their indoor environment's air quality, the hook-and-loop carpet installation system is well suited to maintaining a nontoxic, odor-free installation site. In traditional installation systems, the strong aroma of freshly spread adhesive is often too overwhelming for many people to endure. Often it can take several days for the odor to dissipate. With the hook-and-loop installation system, there are no glues or bonding liquids to produce fumes, so most businesses can resume serving customers as soon as the carpet is installed. Productivity and profits of a business are not compromised because it is off-limits due to harmful vapors. Residential installations also benefit from the "friendly" environmental impact of the hook-and-loop installation system, because of the low toxic emissions. In addition, since the hook tape is reusable for a number of subsequent installations, it reduces future cost and waste. The hook tape can be reused for up to five carpet installations before it has to be changed.

Hook-and-loop versus conventional installations

The introduction of the hook-and-loop carpet installation does not make conventional carpet installation obsolete. Because hook-and-loop products have been on the market only since the early 1990s, it will take a few more years to evaluate how the carpets and installation system hold up over a long period. When more data have been garnered about performance, a true evaluation can be made. Furthermore, the conventional carpet industry, in conjunction with the Carpet and Rug Institute, is making a serious effort to deliver to the public products that are safe, functional, and economical. Also, adhesive manufacturers are producing products that are environmentally safe and less likely to cause long-term health problems to the professional installer. It is certain that both types of installation systems (hook-and-loop and conventional) will continue to be available because each has obvious benefits.

Since many hook-and-loop products often have an attached polyurethane cushion, the cost to manufacture these carpets is much higher than that for a standard polypropylene backing. For example, when you consider the wholesale cost of purchasing a normal apartment replacement carpet, the hook-and-loop products are too expensive. Lesser-quality and lower-priced carpets serve the apartment owner well for short-term use. There will always be a demand for an inexpensive carpet, no matter how advanced technology becomes. Not everyone can afford a top-of-the-line product. The consumer's budget will often be the primary consideration when any floor covering decision is being made.Therefore, from a selling standpoint, it will be necessary to neither oversell nor undersell either type of installation method. It is important to explain the benefits of both products and then let the end user make the final decision. Know when you've said enough. Step back and let the buyers choose for themselves. Be knowledgeable, be honest, and let the consumer make the choice. It's the best way to have a satisfied customer.

7

Floor covering and the environment

As we look to the future, it is important to be aware of what the floor covering industry is doing to protect both the environment and the consumer. Indoor air quality has become a major concern for a great many people and organizations. How flooring products affect individuals who are highly sensitive or allergic to certain chemicals in floor covering materials is one of the primary issues facing manufacturers today.

Among the many innovative steps taken during the past few years toward greater consciousness about these issues, the following programs and enterprises are designed to preserve the environment and safeguard the health of the end user and installer alike:

- Guidelines for the effective removal of materials containing asbestos
- Indoor air quality information for consumers regarding new carpet installations
- Testing program for carpet products to minimize influence on indoor air quality
- The emergence of an industry that reclaims old lumber and old trees for modern-day floor covering use
- The recycling of used carpet, padding, and other soft flooring by-products

Consumer interest in these issues has been heightened by many recent media reports. The flooring industry has responded by establishing stringent manufacturing guidelines and by manufacturing products which strive to diminish the potential health damage. A review of these subjects will give us a glimpse of the present and the future of the flooring industry.

Floor covering and asbestos

Asbestos has been a component in the manufacturing of linoleum floor coverings since the early part of the 20th century. As the years passed and modern medicine became aware of the health hazards associated with the inhalation of asbestos fibers, the use of this raw material in the manufacturing of linoleum ceased. Asbestos was also used in many of the old adhesives, which are no longer available. Even though these products cannot be obtained anymore, their residual effects still linger.

Dealing with flooring materials that contain asbestos is an important issue and one that should be addressed directly. It is necessary to know that there are hazards involved when you handle products containing asbestos. To understand the hazards, it is best to know what asbestos really is.

Asbestos is a mineral mined from the earth. There is evidence that it was used as far back as the Greek and Roman empires. It is one in a group of minerals that have crystal-like structures containing three groups of metal ions. Asbestos is said to be inextinguishable when set on fire. The word *asbestos* actually comes from an ancient term that means inextinguishable. The mineral itself actually separates into long, flexible fibers which were thought to be suitable for use as an incombustible, nonconducting, or chemically resistant material. Asbestos fibers are pliable and soft. They can be spun and woven much like flax, cotton, or wool. Because of this, manufacturers began to combine them with jute and burlap to form the backings for linoleum. Strands were also interlaced with the pressed cork of the wear surface to give the linoleum its strength and stability.

Over the years, up until the late 1960s and early 1970s, a tremendous amount of this linoleum was installed in the United States and around the world. Consequently, it is not unusual for a floor covering contractor to encounter a business or residence with this material installed on the floors. Sometimes this old linoleum is even buried underneath a newer floor covering product. It is, therefore, vital for the person inspecting the job site to discover whether linoleum or any other product containing asbestos is present. Knowledge of the presence of an existing linoleum, or adhesive, that contains asbestos is critical because the inhalation of asbestos fibers found in these floor covering products can cause lung disease or cancer and potentially can lead to death.

It has been proved that the presence of asbestos fibers in the human body can cause irreparable damage. Inhalation of asbestos dust should be avoided. A linoleum or adhesive that contains asbestos should be left undisturbed, if at all practical. It is only when the materials are disturbed, and the fibers are lifted into the air, that they become hazardous. The terms *friable* and *nonfriable* are used to describe this condition.

A flooring product is said to be nonfriable when it is intact in its then-current state. It is considered *intact* when it remains bound to its backing material and the surface has not crumbled or been crushed or sanded. If the floor covering is nonfriable, appropriate measures can be taken to remove the old materials or install new materials over the existing flooring product. Conversely, a product is friable when airborne fibers are emitted into the air. When a friable condition exists, it is dangerous, and specially trained personnel should be called in to handle the situation.

The Resilient Floor Covering Institute (RFCI) has produced a booklet entitled *Recommended Work Practices for the Removal of Resilient Floor Coverings* that outlines effective procedures on how to handle situations where the presence of materials containing asbestos exists. The first item that appears in that brochure is a warning. The warning states that existing floor coverings which may contain asbestos should not be disturbed or tampered with in any way that could create fibrous dust (Fig. 7-1). Anyone involved with the installation of floor coverings should read the RFCI brochure. The information contained in it is quite comprehensive, and the guidelines should be thoroughly examined and adopted.

An understanding of the basic concepts set forth in the RFCI guide is essential. One of the most important warnings is to never sand an existing hard-surface floor covering. You must assume that a product contains asbestos when you are working with old floor covering. Always assume that an old resilient floor covering contains asbestos, unless you have absolute knowledge or evidence to the contrary. You must assume that a product contains asbestos in order to prevent taking an inappropriate action. For example, if you arbitrarily sanded an existing resilient floor covering, only to find out later that it contained asbestos, you could inadvertently expose yourself, and anyone else associated with the project, to dangerous levels of a harmful substance. Therefore, always err on the side of caution.

Whenever possible, it is usually best to put the new floor covering over the old one. Removal of a potentially dangerous floor should be the last resort. To go over an old floor, either you can install a new wood underlayment on top of it (if it is on a wood subfloor), or you

WARNING

Do not sand, dry sweep, dry scrape, drill, saw, beadblast, or mechanically chip or pulverize existing resilient flooring, backing,
lining felt or asphaltic "cut-back" adhesives.

These products may contain either *asbestos fibers* or *crys-
talline silica.*

Avoid creating dust. Inhalation of such dust is a cancer and
respiratory tract hazard.

Unless positively certain that the product is a non-asbestos
containing material, you must presume it contains asbestos. Regulations may require that the material be tested to determine asbestos content.

The RCFI's *Recommended Work Practices for Removal of
Resilient Floor Coverings* are a defined set of instructions which
should be followed if you must remove existing resilient floor
covering structures.

7-1 *Asbestos warning.* (Courtesy of Resilient Floor Covering Institute.)

can cover the existing floor with an embossing leveler and then go
directly to the smoothed surface. If the existing surface is already
smooth, you can install right on it, providing it has been stripped of
all old wax and is clean of any foreign substances.

When any vacuuming of the floor is required, use a *high-efficiency
particulate air* (HEPA) tank type of wet/dry vacuum cleaner. It should
have a disposable air bag and a metal floor attachment. No brush attachment should be used because fibers may escape into the atmosphere through the air spaces between the bristles.

If it is necessary to remove a nonfriable sheet vinyl floor, mix a di-
luted solution of water and dishwashing liquid in a garden sprayer
and wet the surface of the flooring. (The dishwashing liquid should
contain anionic, nonionic, and amphoteric surfactants.) The wetting of
the surface lessens the chances of fibers becoming airborne, thus lim-
iting the potential for them to develop into a breathable component.

Resilient floor tiles can be removed either by spraying them with
water from a garden sprayer or by heating them with a commercial
blow dryer. The heating process helps loosen the tile from the adhesive, which makes removal much easier. Do not dry sweep at any
time during this removal process.

Once the removal of all products is complete, any materials containing asbestos must be put into polyethylene bags that are at least

6 mils thick, labeled with permanent marking inks or with self-adhesive labels, and taken to an authorized disposal center. The label should state that the bag contains asbestos and should not be broken or opened. Thus you help protect others who may come in contact with the bag.

All the above-listed procedures are designed to ensure the safe removal of a very hazardous material. Until all the asbestos-containing materials are taken to landfills and disposed of, their presence will continue to be a relevant issue. Since, however, accepted practice is to either cover up or go on top of a floor covering containing asbestos, the problem will exist for many more years. Ultimately someone will have to deal with the removal. Be hopeful that by that time there may be a better, more practical way to rid the world of this dangerous material.

Indoor air quality

The overall indoor air quality of a given environment can be affected by both carpet and many other household products. Stories of individuals becoming ill after the installation of new carpeting have been reported throughout the nation. Although somewhat sensationalized, the reports do give cause for concern. There are many consumers who are highly allergic to a great number of chemical emissions, and the outgases from carpeting can cause allergic reactions.

Broadloom carpets do emit low-level amounts of chemical offgases. The primary source of these odorous emissions is 4-phenylcyclohexene (4-PC). And 4-PC is present in the latex adhesive that is used to secure the carpet fibers to the primary backing and to attach the secondary backing to the primary backing. Some individuals have allergic reactions or flulike symptoms soon after the installation of new carpeting. Although these are unpleasant and bothersome, only time and adequate ventilation of the premises will alleviate the problem.

The Carpet and Rug Institute (CRI) recommends that windows and doors be opened as much as possible for at least 48 to 72 h after carpet installation. The increased flow of fresh air allows the chemical emissions to be dispelled more quickly. Also, the use of window fans and/or air conditioning units to disperse the emissions to the outdoors will accelerate the process.

The carpet, however, is not always the only source of potentially harmful emissions after a floor covering installation. The carpet cushion or any necessary adhesives could create a noxious odor. Adhesives for a full-spread glue-down carpet installation or adhesives used to secure carpet padding to a concrete floor can contribute to indoor air pollution. Hot-melt seam tape, once heated with an iron, produces fumes that can also have an overpowering effect on some individuals.

All industry manufacturers are well aware of these problems and are consequently producing products that are less harsh and harmful. The reduction in the use of some solvents in the manufacturing process has made current adhesives more *environmentally friendly*. This means that they are safer for both the installer and the consumer. Whenever possible, use products that fit this profile, and safeguard against air pollution by providing proper ventilation and fresh air circulation at the job site.

A number of consumers are concerned about the use of *formaldehyde* in the production of broadloom carpets. Formaldehyde, although used many years ago, is no longer a part of the manufacturing process today. Some people have extreme adverse reactions to products that contain formaldehyde. Knowing that it is not used in manufacturing anymore, you can assure clients that new carpets do not contain this harmful chemical.

Dust is also a factor contributing to poor indoor air quality. When old carpet is being removed, the dust embedded in it is stirred up and circulated into the air. Sensitive individuals may be affected by this occurrence. Vacuuming the carpet prior to its removal will minimize the dust particles still in the carpet at the time of its removal. Also, whenever possible, vacuum the subfloor after the old carpet has been removed to clear away any dusty residue. After the new carpet is installed, vacuum it thoroughly to remove any loose fibers. From then on, tell the consumer to establish a regular maintenance program to minimize dust and dirt buildup. The more often a carpet is vacuumed, the less likely it is to contribute to poor indoor air quality.

For individuals who are extremely sensitive to the by-products of chemicals used in the manufacture or installation of floor coverings, the best advice you can give them is to be absent from the premises while the old flooring is being removed and the new flooring is being installed. If necessary, suggest that they stay away until the odors from the chemicals have completely dissipated, which may be several days. When possible, plan installations for a time when use of the area will be minimal.

Indoor air quality
carpet testing program

The carpet industry, in conjunction with the CRI, has voluntarily instituted a research program to test for certain chemical emissions from broadloom carpets. The CRI has developed a certification program, widely referred to as the *green–label* program, to ensure that the carpet has been tested and found to be free of harmful substances. The green-label logo (Fig. 7-2) is attached to all carpet samples that passed the tests with satisfactory results. The label has a product type code imprinted on it. This code consists of a series of numbers and/or letters (for example, 1Y24030729) that refer to a certain product type manufactured by a specific supplier. The product type denotes the kind of backing material, fiber content, chemical makeup, and dye process used. The use of the green-label logo is restricted to participating manufacturers who have actually had that specific product tested by an independent laboratory.

Programs such as this will bring about technical advancements that may make carpeting even safer in the future. Lower chemical emissions and a reduced impact on indoor air quality are indeed a goal worth striving for regarding flooring products.

Reclaimed woods

An emerging industry in the floor covering field is the retrieval and remilling of previously used lumber. This new category of products is called *reclaimed,* or *antique, woods*. While the majority of this wood is pine, other possibilities are walnut, hemlock, chestnut, and poplar. The pine that is typically found is longleaf pine, sometimes called heart pine. When the United States was being colonized in the early 1700s, forests of longleaf pine had been standing for centuries. When the industrial revolution began in the 1800s, many of the factories and warehouses of eastern cities were built of longleaf pine. Furthermore, many of the barns in the whole eastern coast of the United States were built using this same material (Fig. 7-3). Today, as many of these structures have become run-down and dilapidated, certain visionaries have seen the value in these old timbers. In fact, some logs are even being rescued from river bottoms, or deep cold lake bottoms, where they have lain for over three-quarters of a century. The retrieval process, though slow and excruciatingly difficult, has yielded lumber that can be reprocessed into an amazingly beautiful floor.

Boards, or beams, taken from any of the aforementioned locations are shipped to the salvager's mill where they are inspected for possible future use. Only about half of the lumber retrieved will be usable. If it is qualified for reuse, it will be resawed and dried. Even though these materials are old, they are still susceptible to the same fluctuations in moisture content as all woods. Although the sap content has evaporated, to ensure dimensional stability and minimal

INDOOR AIR QUALITY CARPET TESTING PROGRAM

FOR MORE INFORMATION
The Carpet and Rug Institute
1 - 800 / 882-8846

product type:

INDOOR AIR QUALITY CONSUMER INFORMATION

IMPORTANT HEALTH INFORMATION:

SOME PEOPLE EXPERIENCE ALLERGIC OR FLU-LIKE SYMPTOMS, HEADACHES, OR RESPIRATORY PROBLEMS WHICH THEY ASSOCIATE WITH THE INSTALLATION, CLEANING, OR REMOVAL OF CARPET OR OTHER INTERIOR RENOVATION MATERIALS. IF THESE OR OTHER SYMPTOMS OCCUR, NOTIFY YOUR PHYSICIAN OF THE SYMPTOMS AND ALL MATERIALS INVOLVED.

SENSITIVE INDIVIDUALS:

PERSONS WHO ARE ALLERGY-PRONE OR SENSITIVE TO ODORS OR CHEMICALS SHOULD AVOID THE AREA OR LEAVE THE PREMISES WHEN THESE MATERIALS ARE BEING INSTALLED OR REMOVED.

NOTE:

YOU CAN REDUCE YOUR EXPOSURE TO MOST CHEMICAL EMISSIONS WHEN CARPETS AND OTHER INTERIOR RENOVATING MATERIALS ARE INSTALLED, CLEANED, OR REMOVED BY INCREASING THE AMOUNT OF FRESH AIR VENTILATION FOR AT LEAST 72 HOURS. (See Installation and Maintenance Guidelines or ask for Owner's Manual.)

INSTALLATION GUIDELINES:

- VACUUM OLD CARPET BEFORE REMOVAL

- VACUUM FLOOR AFTER CARPET AND PAD HAVE BEEN REMOVED

- ALWAYS VENTILATE WITH FRESH AIR (OPEN DOORS AND/OR WINDOWS, USE EXHAUST FANS, ETC.) DURING ALL PHASES OF INSTALLATION AND FOR AT LEAST 72 HOURS THEREAFTER

- IF ADHESIVES AND/OR PAD ARE USED, REQUEST THOSE WHICH HAVE LOW CHEMICAL EMISSIONS

- FOLLOW DETAILED INSTALLATION GUIDELINES FROM MANUFACTURER OR FROM CARPET AND RUG INSTITUTE

INDOOR AIR QUALITY
CARPET
TESTING PROGRAM
product type

800/882-8846

The manufacturer of this carpet participates in a program which seeks to develop ways to reduce emissions by testing samples of carpet. With fresh air ventilation, most carpet emissions are substantially reduced within 48-72 hours after installation.

FOR MORE INFORMATION: CARPET AND RUG INSTITUTE 800/882-8846

7-2 *Indoor air quality carpet testing label and information.* (Courtesy of Carpet and Rug Institute.)

7-3 *Old barn made of longleaf pine.* (Courtesy of Aged Woods, Inc.)

shrinkage, the boards should be brought to the same moisture content level applicable to new boards, which is 6 to 9 percent. Also, like any other wood product, these boards should be brought to the final job site many days before installation to allow the wood to acclimate to those atmospheric conditions.

Prior to being resawed at the mill, the boards are run through metal detectors to locate any nails or other foreign objects. The longleaf pine is then graded by a system that is unique to each company. Since the wood is a reclaimed product and not so readily identifiable like modern woods, recognized grading standards (such as those of NOFMA) do not apply to these woods. However, to arrive at some semblance of a grading system, the following factors are taken into consideration: the amount of heartwood versus sapwood, density of the grain, knot content, grain pattern, weather cracks, and the amount of nail holes and/or worm holes.

Old-growth pine extracted from the virgin forest of the southern United States is more durable than modern pine because the old trees had a tighter grain pattern. Trees that are only 40 to 50 years old do not have the grain density of a 150-year-old tree, and it is grain density that determines the hardness of a wood. Since these old forests are virtually gone now, the only way to enjoy the exquisite grain pattern and texture of those old woods is to reclaim them from existing structures before they are torn down and hauled off to landfills as scrap. If that happens, they will be lost forever. The richness and warmth of these wood products far surpass those harvested today. Not only are they unique, but they all come with a little bit of history. Whether they were part of a barn built before the Civil War or came from logs found on the bottom of a lake, they are remnants of our cultural past. Remilled and put into a contemporary structure, they make a beautiful contribution to any building. As architects and designers continue to discover new uses for these sleeping beauties, the reemergence of these long-forgotten gems is ensured. Specified for residential as well as commercial use, these products can be used as floors, walls, and ceilings. Moreover, since they are reclaimed woods, their use reduces the need for a growing tree to be cut down.

Inasmuch as the natural forests and rain forests are being depleted at an alarming rate, any form of wood recycling is a welcome relief. Unfortunately, however, reclaimed woods are a limited resource with a finite capacity; once they are gone, they are gone forever. Furthermore, they could never meet today's growing demand for wood products. Their exclusivity is protected by the slightly higher prices set for the purchase of these specialty woods. Yet, it is good to know that these valuable woods are not simply being discarded as useless leftovers of a bygone era.

Recycling in the flooring industry

Recycling centers have been established throughout the United States to collect used carpet and padding. These centers, mostly established by industry manufacturing companies in response to the obvious need to dispose of discarded floor coverings, are a step toward responsible resource management by the floor covering industry. Individuals concerned with the piles of trash in community landfill areas are pressing for a reasonable way to dispose of used carpet, padding, tiles, solvents, and adhesives. At one carpet recycling center, standard procedure is to begin by separating the backing from the fibers. The fibers are then treated and reconverted to a reusable polymer, which can then be made back into carpet fibers. These fibers are indistinguishable in appearance and quality from new fiber. Waste plastics, fibers, and polypropylene backings are distributed, or sold, to other industries. Companies in these other industries make a wide variety of products from these materials. This program, however, is a new one, and it may be several years before it is in force in most communities.

Another unusual way in which the carpet industry is participating in a large recycling effort is through the purchase of recycled plastic beverage containers. These are being processed into polyester fibers which, in turn, are manufactured into carpet fibers. Polyester carpet fibers are highly fade-resistant, and they readily accept dyes for vivid color reproduction.

Programs that preserve our natural surroundings are needed so that future generations will inhabit a world that is equal to, or better than, our world today. Responsible management of our natural resources and raw materials, as well as the proper disposal of waste products, will ensure a more balanced environment. The challenges that lie ahead of us are mammoth, but not insurmountable.

8

Floor covering as a business

Floor covering, like any other business, is just that—a *business*. To be successful, you must approach it with confidence, dedication, and a well-thought-out plan. Now that you are "armed" with the knowledge contained in the preceding chapters of this book, the next logical step is to establish yourself as a participant in this profession.

It has been assumed from the beginning of this book that you are already involved in the building trade in one form or another. Whether you are a floor covering installer, a general contractor, or a specialty contractor, your interest in floor covering as an extension of an existing business or as a new venture is the premise upon which this book was written.

Far too often, tradespeople who desire to open a business of their own lack the knowledge and guidance in how to make their ideas become reality. A well-informed individual will have a greater chance of success than someone who proceeds by trial and error alone. The ultimate objective of anyone entering a given field should be to acquire enough information on the subject to make that transition as effortless as possible.

All successful businesses are founded upon sound, fundamental principles. The following sections explore numerous ways you can begin a business, finance it properly, and operate it profitably. Effective decision making is possible once you have a roadmap that will show you such strategies as purchasing, marketing, sales, and advertising. Countless other details are contained in these sections and give an overview of how to start a prosperous floor covering operation.

Entrepreneur or employee

An entrepreneur has often been defined as an individual who can organize, manage, and assume the risks of a business or an enterprise. To do this, one must have the ability, as well as the resolve, to make the business successful. Therefore, the first question you must ask yourself is, Do I have what it takes to be an entrepreneur? If the answer is yes, then you have cleared the first hurdle on your way toward career fulfillment, since you have the confidence to believe in yourself. If you were to lack the confidence in your own abilities, any one of the numerous obstacles you will encounter in getting started could easily cause you to contemplate giving up long before you should.

Many people think about starting out on their own because they find their present employment situation lacking. Often, it's not that they are not making enough money; it's because their current job does not give them the creative outlet they most desperately desire. The personal satisfaction enjoyed when one's career mirrors one's skills and ambition is unparalleled. The freedom to control one's destiny is part of the "American dream." Yet that freedom carries with it grave responsibilities—to yourself, your family, and your company. All those individuals involved with you daily now become dependent upon you. The decisions you make could very well affect them personally. Therefore, take this responsibility seriously before you embark upon any business venture, no matter how simple it may seem from the outset. Always remember that no matter how powerful or successful you may be in the future, you will have one absolute boss—your customer.

Customers are the lifeblood of nearly every business. Without them, a business ceases to exist. If you are dedicated to providing excellent customer service, your chances of success are greatly enhanced. Too often, business owners take their customers for granted and lose sight of the fact that without customers there are no resources to fund the payroll or other ongoing expenses. Consequently, entrepreneurs must be ever-mindful of their customers' degree of satisfaction.

Another reason why people want to start their own businesses is the unlimited income potential. In light of the massive corporate downsizing taking place in the workplace today, there is very little job security. We are all, therefore, responsible for our own future. A person with entrepreneurial skills knows this fact and will take the steps necessary to achieve their financial independence.

Recognition and prestige are yet two more reasons why people go into their own business. Undoubtedly, one enjoys a certain amount of renown in their community when they become a business owner. Any successful business that has been operating for any length of time is a recognizable institution in that locale. With that recognition comes the prestige that opens the doors to power and influence. Invitations to community-sponsored events, and your participation in those events, will enhance your standing in your local environment and at the same time afford you an opportunity to give something back to the people who helped make you a success.

Perhaps the most important reason some people choose to go into business for themselves is *the challenge*. The challenge of pitting one's resources and skill against the odds of failure can be a constant source of inspiration and adrenaline. While a normal job may provide for your personal needs, the pride you develop by striking out on your own is incalculable. This satisfaction can often make up for the long hours and stress that will accompany your new position.

The challenges that lie ahead are formidable, but not insurmountable. Most businesses fail because of poor management. Therefore, it is incumbent upon you to prepare yourself adequately against the common pitfalls encountered by most businesses. By developing good management skills you can increase your chances of success. The location of your business, the source of prospects for your product, and the strength of your competition should all be assessed well in advance. To properly address these issues, first compose a written business plan. A company that lacks such a plan may soon flounder and become nonexistent.

Business plan

A formal business plan is much like a course that a ship's captain charts before leaving the harbor. No sane captain would ever consider setting sail over a large expanse of water without first consulting the appropriate maps and tide tables. Composing a business plan is nothing more than stating the company's mission and establishing goals to help achieve that objective. By taking the time to set forth these goals, in writing, you will have better insight into what you want to accomplish and the steps you will need to take to get there. Moreover, these guidelines are especially useful during times of crisis to help keep your company on its intended course.

A well-devised plan helps focus the necessary energy and time on those facets of the business that are critical to its survival. It helps eliminate wasteful activity that is not essential to the company's success. Therefore, all efforts can be directed toward areas that can be most profitable.

A business plan can be as broad or as narrow as you wish, but the critical feature in regard to its usefulness is that it must be *in writing*. The act of writing it on paper gives it a sense of permanence and significance. It also gives you something you can refer to at a later date to check your progress.

In addition to writing down the goals you want to achieve, you should list the obstacles you will need to overcome. Here again, by writing them down and examining them periodically, you will realize that they are less significant than you originally thought and that the means necessary to overcome them are well within your grasp.

Your business plan should include short-term goals as well as long-range objectives. It should describe your market, competition, and sales projections. It should also outline your strategies for marketing, advertising, and other operating procedures. It is important to set a specific timetable for the completion of these goals. By not setting a date for their completion, you run the risk of never accomplishing them. By specifying a certain time frame for their attainment, you inject a kind of immediacy and urgency into the process. That is not to say that there needs to be a sense of stress or panic. It just means there needs to be a deep-seated desire to unhesitatingly begin to work toward these goals. In addition to all the above-stated goals and strategies, there must also be a financial plan prepared.

Your financial plan should explain where the money for your business is coming from. Whether it is coming from your personal savings, investors, relatives, or the sale of stock, all this should be spelled out in black and white. You should draft a budget list that includes all start-up costs and fixed expenses. Combine that with anticipated normal operating expenses, and you will have a pretty good idea of how much this venture is going to cost you. Then estimate your projected sales figures based on all the research you have done regarding this endeavor, and make an educated guess of your profit potential. If your profit estimates do not greatly outweigh your expenses, now is the best time to reevaluate the entire concept.

Over the years, several studies have shown that individuals and companies that prepare viable, written business plans are far more successful than those who do not. To improve your bottom line, follow the lead of these prosperous enterprises and devise a business plan of your own—it can only be to your benefit. Evaluating your strengths and weaknesses will help you rise to the challenges ahead with confidence and clear thinking.

Selecting the best location for your business

You have undoubtedly heard, countless times, the old adage that the three most important requirements for a successful business are "location, location, location." Well, as mundane and as tired as the saying may sound, the truth of these remarks is undeniable. New business owners often make the mistake of choosing an out-of-the-way location simply because they were able to obtain it at a low rent. Yet, once they open the business and discover that they are having a difficult time attracting customers, they realize that they have to expand their advertising campaign to draw in clients. Consequently, they wind up spending about as much in rent and advertising as a store in a better location would be paying for rent alone. A store that has a good location can often get by with a little less advertising because the site is visible to such a large number of potential customers. If the product that the business is selling is desirable enough, those people passing by will ultimately come in; or when the product is needed, people will remember the business because of its visibility. This is not to say that you should not search for a location at a reasonable rate, or that you should choose the most noticeable place in town. But you should weigh all the factors before making a decision based solely on a monthly rental price.

By the same token, we are talking about a floor covering business location, and not a clothing store location. There is a big difference between the two. First and foremost, the floor covering business is not a walk-by type of business. People do not just wander into the local floor covering store and browse as they do for dresses and slacks. They come in the store because they *need* your product. They are out shopping for carpet. It's as if they are on a mission. They will, more than likely, make a conscious decision that they need carpet, and then they will make a concerted effort to procure it. They will probably pick up the telephone book; write down the names and addresses of three, four, or five flooring companies; and drive to each to see what is available. Based on that search, they will decide where to purchase their flooring. With this information in mind, we need to look at the special requirements of a floor covering store location compared to those for most "normal" retail operations.

Before you even begin to look for a location, determine whether you are going to buy a building, build a new building, or rent a space.

If at all possible, it's generally best to purchase your business location. There are several distinct advantages associated with this option:

- You need not worry about losing your location.
- There are no unanticipated rent increases.
- There are certain tax benefits.
- You can always mortgage the property to raise needed cash.
- You can sell the property and often make a sizable profit.

One of the decided disadvantages of buying a piece of property is the high initial cash outlay. However, when you own your location, all your labor is going toward building your own future, not the future of your landlord.

If you choose to build a building, one advantage is that the design will be to your specifications, and so you will have exactly what you need and want in a structure. Considerable time, energy, and money will be needed to bring this about, so you must ensure that building the premises is the right choice for you.

Many businesses rent a location, and for many people it works out fine. One of the prime reasons most businesses start out by renting a facility is because the initial cash investment is relatively low. A typical rental agreement calls for a month or two of rent in advance as well as a security deposit. If you are operating with limited resources, this leaves the majority of your finances intact for other opening expenses. However, a major drawback of renting is that you are vulnerable to rent increases or could lose the location you've worked so hard to develop once the lease runs out. But, assuming you are going to rent your facilities, here's how to go about it.

Renting a business location

Begin your search for a business location by sitting down calmly with pen and paper to determine the necessary features for your ideal structure. Since a floor covering operation has certain minimum requirements, facilities that lack these features should be eliminated. Some of the required items are as follows:

- Large overhead garage-type rear door. Entrance through that door should be via a ramped concrete floor to sustain the weight of a forklift and/or other heavy equipment.
- Concrete floor (at least in the warehouse area) to support the weight of the inventory and operating equipment.
- Large enough building to have a showroom and a warehouse in the same structure. This gives the owner greater control and the ability to oversee the entire operation by having everything combined in one location.

- Parking. If at all possible, providing adequate parking for customers is a definite plus.
- Easy access for deliveries from semitractor/trailer trucks.
- Visibility to those passing by in vehicles or as pedestrians.
- High ceilings to accommodate stacking of inventory and the raising and lowering of a forklift mast.

As you contemplate these and other possible features, try to conceive of the overall image you want your company to have. This image will help you choose the right space.

After you have written down these general needs, study a map of the area where you would like to start your business, and determine what section of town would best suit your enterprise. Then check the local papers and real estate advertisements to see whether anything is available in the area that fits your requirements. If appropriate, contact several real estate agents to inquire about leasehold availability. More importantly, drive to this section of town and inspect it personally. Quite often a building may have a "For lease" sign on or near the structure but may not be advertised in the newspaper or be listed with a real estate agent. Some property owners prefer to have their buildings shown in this way.

Once you have spotted an interesting possibility, contact the responsible party so you can inspect the interior. Check out everything during your walk-through; if you are really interested, ask permission to bring a general contractor with you to inspect it at a later date.

First, ask the rental price per month. Then ask how much must be paid upon the signing of the lease and whether a security deposit is required. Next, ask the length of the term of the lease and whether there are any option years. Find out what is included with the rental price (i.e., garbage disposal costs, repairs, insurance, utilities, taxes). When you get the answers to these questions, compare these figures to the amount you had budgeted for rent. If the dollar figure and the location and structure are in line with what you had preplanned, continue to gather as much information as you can about the property so you can negotiate properly. Remember, everything is negotiable. The more things you can include in the original lease that are to your benefit, the better off you will be. Once the lease is signed, you're stuck with every detail it contains, good and bad.

Check out the surrounding neighborhood, and talk to any other tenants, if possible. Check the zoning codes to make sure your business will not violate those laws. Contact the local authorities to verify signage allowances and restrictions. Make certain the building meets code requirements for the electrical, plumbing, and/or other physical elements.

Also, is the landlord willing to help you pay for any remodeling expenses needed for you to bring the property to a level where you can open a business? If not actual cash, many building owners will give a month or two of free rent to help compensate you for any repairs and renovations that may be necessary. Make concessions, and offer to share some of the expenses if you see the landlord is willing to try to fulfill your requirements. Bargain whenever you can, and offer a little less than you are willing to pay. You never know, the landlord might accept it!

After you have hammered out all the details, put everything in writing. Read it over carefully, and have an attorney and/or real estate agent read it as well. Once you have comprehended all the issues set forth in the document, take some time to think about it. Sleep on it. Let your mind and heart both agree that it's the right thing to do— that it's the best location for you at the best price and terms. When you are sure that you are making the correct decision, sign the lease and begin your career as a promising entrepreneur.

Purchasing an existing business

For some people, starting a business from nothing is a frightening proposition. By purchasing a going concern, the transition from employee to business owner is made much easier. Keep in mind that the purchase of a business must be a wise one, so there are a great many factors to consider before you make that purchase. If you thoroughly investigate the company and find it to be viable, it may be a very good move. If you pay the right price and the company has a good reputation and plenty of potential, under your guidance the business could be making money in no time.

Just as you have your reasons for wanting to buy an established business, the seller will have motivating reasons as well. Not everyone wishing to sell a business is doing so for deceptive purposes, but you should be aware of that possibility. Owners often merely wish to leave the company because of health reasons, they may want to retire or may be simply burned out from doing the same thing for a great number of years. At any rate, ask why the owner has decided to sell the business. Investigate the premises with an open mind, and ask yourself whether you believe the current owner's answer to be the real reason for the sale. It is your responsibility to properly evaluate the physical and financial conditions of the company, so research every aspect of the enterprise carefully.

There are considerable advantages to buying an existing business:

- The location, equipment, fixtures, and inventory are currently in place.
- There is no need to go through the painful start-up phase.
- There is an established customer base that you can draw from immediately.
- Generally there will be no, or very little, renovation costs.
- If the business has been successfully established for over 5 years, it has a good likelihood of continuing to be successful.
- Owner financing may be available.
- The current owner may stay on for a while to act as a consultant to make the transition simpler.

Sometimes you will find a business that is doing only marginally well. In these instances, reviewing all the circumstances involved, you may find that the reason for its lack of success was poor management, poor location, competition, or other specific causes. With your unique style and dedication, you may be able to overcome these obstacles and change the direction of the business. This type of business opportunity may be just right for you.

When you buy an established business, often you will be paying some form of remuneration for what's known as *goodwill*. Goodwill is an intangible asset. Unlike a tangible asset which you can usually touch, such as equipment, furniture, supplies, and inventory, the value of goodwill is based on elements that greatly contribute to the expected profits of the company. Intangible assets include a firm's customer base, patents, trademarks, and managerial style. Placing a value on goodwill is never a clear issue, but you should expect to compensate the previous owner for all the hard work contributed over the years if the company is reputable and will bring clients to your doors immediately. There is a lot to be said for walking into a business on your first day as owner to find the phones ringing off the hook with customers placing orders or calling to schedule home measurement appointments. Since a quick return on your investment is what you are hoping for, you must give the prior owner her or his due for allowing you the ability to be able to accomplish that feat quite quickly.

By the same token, examine that customer base. Make sure you will not have any difficulty maintaining their support. If their loyalty is closely bound to the prior owner, you may have a difficult time persuading them to be faithful to you. If you feel that the customers will support you as well as the previous owner and if you can use that base of clients to create an even wider range of customers, then the money you spend on goodwill may be worthwhile. It is the existing customer base that will help you meet expenses and pay for advertising to help expand your business.

Another important factor to consider is the element of current employees. Certain key employees may be critical to your early success. If one, or several, of those employees were to leave the company soon after the sale, it could disrupt your plans and hurt you financially. On the other hand, there could be certain employees who are influential in operating the company, yet are detrimental to it. In those cases you may find it necessary to rid the company of their services immediately, which will require you to evaluate, hire, and train your own staff. A disruptive employee can severely handicap your initial efforts and cause you untold damage. It's often best to start fresh and create your own approach to the company.

A person selling you a business will often include, as a part of the deal, consulting services for a certain period. It's not unusual to have that person on hand for a month or more to walk the new owner through all the steps of the business. The seller may also stay on for several months on an on-call basis to advise the new owner, as unexpected circumstances arise. This can be a valuable assistance. In the first month or so, you may be so overwhelmed by the complexity of it all that you won't be able to anticipate all the potential problems. Having someone you can call to help you through these difficult times is worth a lot. To help you overcome an installation problem, or an accounting difficulty, for example, the previous owner's experience may assist you and give you great peace of mind.

When you examine the financial records of a company, always consult an attorney or an accountant. Let them help you evaluate the records and assist you as you negotiate the transaction. They will know what to look for in the files that will give you a true indication of the financial position of the company. Undoubtedly it's desirable to buy a business that's showing a huge profit, but you will pay for that dearly. A business that shows a loss in one year but profit in others could be explained by several factors. Capital improvements could have been made, or certain writeoffs could have been taken during that year. Perhaps payments currently being made on certain debts will be paid off by the owner upon the sale of the business. Consequently, those loan and interest payments will no longer cut into the new profits. Also certain other expenses may exist now that won't be there once you take over. By trimming unnecessary expenses you can strengthen your company's financial position rapidly.

Examine current profit-and-loss statements, and tax returns for at least 3 years, very thoroughly. Go over them with your advisers to get a clear picture of the operation. Take all their advice and concern to heart; but remember, in the final analysis, it is your decision alone that will matter. Keep your own counsel, and make the choice that best suits your purposes. If you have pervading thoughts of disaster, back away and look for another opportunity. But if you feel confident that this is the right thing to do, give it a go. Weigh the risks, but don't make a risky decision. Base all your conclusions on sound fundamental facts, not pie-in-the-sky speculation. If you have to hope for the nearly impossible to happen to be successful, it's probably a bad bet. But if the odds are in your favor, the ability to come to an intelligent decision and act upon it is what separates the winners from the losers.

Equipment

The most important piece of operating equipment required for any floor covering operation is a forklift (Fig. 8-1). The forklift should have a "rug-pole" attached to its lifting mechanism instead of actual forks. This makes it easy to move a roll of carpet. Never consider opening a store that sells carpet without first purchasing a forklift; it is indispensable to your enterprise. Some large carpet rolls can weigh up to or beyond 1500 lb. Wrestling with rolls this size is nearly impossible, no matter how many hands are at work. Furthermore, moving these rolls daily soon takes it toll on a person's physical well-being. Chronic bad backs and strained muscles are commonplace in

8-1 *Forklift with carpet pole attachment.* (Courtesy of Container Bins, Inc.)

the flooring industry. These problems intensify when the store owner refuses to invest in a piece of equipment that can only benefit every-one concerned. Simply put, if you cannot afford to purchase or lease a forklift, reconsider the floor covering business. In the long run, your health, or the health of your employees, will begin to suffer; and all the regret in the world won't return that health. Furthermore, you may find yourself shying away from bidding for larger jobs because you will be fearful of handling such heavy rolls of carpet; right there you're digging into your bottom line. So include a forklift in your financial business plan. It doesn't have to be a new one. Some of the best values are reconditioned models, and they are widely available at a reasonable price. Check your local telephone book for dealers in your area so you can compare prices and features.

There are three extremely important features to look for when shopping for a forklift:

1. Free-lift
2. Side-shift
3. Tilt

All three features are associated with the mast of the forklift. The mast is the part of the forklift that goes up and down. The rug pole is attached to a mounting mechanism on the mast. The mast comprises a heavy metal, permanent shaft. Inside this shaft are steel cylinders that raise and lower within and above that mast. *Free-lift* provides for the raising of the rug pole several feet before the cylinders start to rise above the upper levels of the permanent mast. By keeping the cylinders below the upper level of the permanent mast, you can operate it in low quarters and still lift up a roll of carpet. *Side-shift,* as the name implies, allows the rug pole to be shifted from side to side without having to move the vehicle itself. This is the most important feature in a forklift, because the angle of approach to a roll of carpet is vital. A roll of carpet is wrapped around a stiff cardboard tube. If you penetrate the tube at an improper angle, you run the risk of puncturing the tube and tearing the carpet. There is nothing worse in this business than getting a call from your installer and being told that the last 10 ft of a custom-ordered roll of carpet is ruined and use-less because the forklift pole tore through the back of the carpet in several places. The side-shift feature gives you tremendous lateral range when lining up your entrance into a carpet tube. The *tilt* fea-ture adjusts the mast in a forward and backward manner. Like the side-shift, it gives you greater ability to line up your approach and enter the tube on a level angle.

The necessity of having a forklift cannot be overstated. The right one for you will be largely determined by the specific characteristics of your building. When you shop for a forklift, you may consider some of the other main features:

1. *Fuel system.* Forklifts can typically be obtained that operate on gasoline, liquid propane gas (LPG), or electric power. (Of the three types, the LPG is perhaps the best.) Using a gasoline-operated forklift in the confined quarters of a showroom/warehouse can cause the release of harmful toxins into an otherwise clean working environment. A gasoline-fueled forklift is best suited for outdoor use. The electric battery-powered system, while more hygienic in its emissions output, is more unpredictable in its usability. If, for example, the battery is not properly charged, the forklift simply won't work. Far too often, someone may forget to plug in the battery the night before a big job is to go out, only to have the installer arrive at the warehouse the following morning to find the forklift out of commission. With no other way to lift a heavy roll, it will take hours before the system is properly charged. Running out of "juice" is the biggest drawback to an electric forklift. Invariably it seems to fail just when you need it most. Forklifts that operate on LPG are less toxic than the gasoline models, have plenty of power, and are very dependable. By keeping an eye on the fuel gauge, you can be assured of a reliable, high-performing piece of equipment whenever you need it.

2. *Overall dimensions.* If you are operating in confined quarters, be sure to check the overall dimensions of the vehicle you are considering. If you have a narrow or low door opening to your building, make sure the forklift will fit into the opening.

3. *Maintenance accessibility.* Check the access to the engine compartment in case minor repairs and maintenance are needed. Although major work should be done by a qualified mechanic, if your engine is readily accessible, you can do the small repairs yourself. Some forklifts are covered by a metal housing that prevents a layperson from even attempting the most minute adjustments. These forklifts have to be hauled, via flatbed trailer, to the service center. If the servicing company does not give you a loaner vehicle, you could be without a forklift for days. Couple this inconvenience with the high cost of transportation and repairs, and this is a major expense. So try to locate a vehicle that has easy access to its engine components.

Another piece of equipment that may be necessary is a pickup or flatbed truck. Having a vehicle on hand that can haul several rolls of carpet, pad or any number of items is crucial. Some flooring shops can get by utilizing the installers' vehicles, but it is handy to have your own form of conveyance. It enables you to make customer deliveries or pick up supplies at the supply shop when needed. However, as with any major capital expense, you have to consider whether it is better to buy or lease the vehicle.

There are pros and cons to buying or leasing a forklift or truck. When you buy a vehicle, you own it outright and are responsible for all the maintenance. Whenever you want, you can sell it and purchase another. How you take care of it and its condition when you put it up for sale will determine the price you get for it. When you lease a vehicle, however, there are more factors to consider.

When you lease a vehicle, one good feature is that there is typically a minimal amount of out-of-pocket cash needed. This frees up a lot of your initial working capital to help get your business off the ground. Those individuals who spend too much money purchasing expensive items before they've generated even one dollar of revenue for their new venture often find themselves in serious financial trouble from the very beginning. Therefore, whether you purchase or lease a vehicle, keep your initial down payment to a minimum.

While there may be certain tax advantages to leasing a vehicle, one of the main disadvantages is that no equity is gained. Carefully consider the terms of the lease before you decide on this mode of acquisition. Check how many miles you will be allowed to put on the vehicle before any penalties are incurred, and find out what will happen if you have to terminate the contract prematurely. Get as much advice as you can from a tax consultant or leasing expert before you commit to any lease agreement.

Regardless of the method you choose for acquiring the product, remember to separate your needs from your wants. If a $12,000 vehicle will suffice for the work you have planned, don't buy a $22,000 model. It's a needless expense that you can ill afford at this stage of your enterprise. Don't be too eager to part with your capital. When you purchase any item for your business, say to yourself: "If it's going to make me more money, then it's the right decision; if not, then forget it."

Choosing a legal structure for your business

Before you can begin operating your enterprise as a viable concern, you must choose a legal form for your business. Basically there are three legal ownership options:

- Sole proprietorship
- Partnership
- Corporation

Sole proprietorship

The sole proprietorship is perhaps the simplest and most widely used form of ownership in the United States today. It is the easiest of all the three forms to establish. As a sole proprietor, you are the singular owner of your business. All your personal funds and your business funds are legally considered to be one. Therefore, if your business is unsuccessful and the company assets are insufficient to pay off your obligations, your creditors can attach your home, take away your car, or remove money from your bank account to satisfy the debts.

To create a business as a sole proprietorship, the setup costs are very low if you plan to operate it under your own name. If, however, you decide to use a trade name, you have to file a form with the local governing agents called a *fictitious business statement*. The cost for filing this statement is nominal. In addition, usually the information contained in the statement must be published in a local newspaper. This requirement legally notifies the general public that you are now conducting business under an assumed name. Anyone who has any objection to your operating under that name must contact the authorities as soon as possible to prevent you from doing so. Since fictitious business statements are required to be published once a week for several consecutive weeks, when that time is over, you can use that business name for the duration of the agreement without fear of repercussions.

In a sole proprietorship you are the boss. You alone are the one who will be recognized for the success of the business or ridiculed for its failure. You are the only one who will reap the profits or suffer its losses. If the company is sued, you are sued. From a tax standpoint, any loss or gain from the company goes directly to your personal income tax statement. When you are ready to terminate the business, you simply pay off the debts and close up shop. There is no other person or entity to consult.

Starting a business as a sole proprietor affords you a low-cost way of beginning your company. Still, if at a later date you with to incorporate or take on a partner, there is nothing to prevent you from doing so. You are free to change anytime, providing it is done within the framework of the law.

Partnership

A partnership is an agreement (preferably in writing) between two or more individuals for the purpose of conducting a business. These individuals are commonly referred to as *principals*. A partnership is very much like a sole proprietorship except there is more than one person to share the responsibilities. The partners you choose may be involved in the day-to-day operations, or they may be what's referred to as "silent" partners. This type of partner is usually just an investor in the business who plays a minor role in its management.

In an ordinary partnership, where two individuals are involved, each person has a 50 percent share. Each partner, known as a *general partner,* is jointly and equally liable for all the activities of the business. Both individuals are personally liable for the debts of the company. A partner's personal assets can be seized to pay any business debts. Also, any action or business decision made by one partner legally binds the other partner; this can apply even if one partner knew nothing of the circumstances.

Another type of partnership agreement is a *limited partnership*. In this type of business arrangement, there is a general partner (or partners) as well as one or more "limited" partners. The limited partners invest in a company and have the possibility of sharing in its profits. However, if the company were to lose money, the liability of limited partners is only up to the amount of their investment. The general partners are personally liable for the firm's debt and also are responsible for the ongoing management of the operation.

On the positive side, a partnership agreement is good because it brings additional skills and capital to the venture. In some businesses the workload is so enormous that one person would have a difficult time controlling it all. Also, one's personal financial resources may be insufficient to fund a new or growing concern. The attributes one partner brings to the deal may very well complement the attributes of the other. If the individuals involved are compatible and share the same goals, it could be a match made in heaven; if not, it could be one of the worst decisions of a lifetime.

Just as with marriage or living with another human being, compatibility is the key component of any partnership. Without compatibility, a partnership is a sinking ship. The ability to cooperate and communicate with your partner is an essential element in having a successful partnership. For someone who is a sole proprietor, the decision to take on a partner can result in a difficult transition. Far too often, partnerships are begun with the best intentions, but personal incompatibility turns them into a nightmare. It's often difficult enough for married people to get along, let alone virtual strangers. Therefore, knowing the person you are about to go into business with is very important. Getting involved with people just because they have money to invest can sometimes be the wrong reason. Then again, forming a partnership with close friends or relatives, without carefully evaluating their potential in your business, can be equally disastrous.

The relationship you form with your partner could determine your emotional well-being for many years to come. Disagreements can arise over major issues as well as trivial matters. How you respond to these situations will form the basis of your association with each other. Personality clashes are not uncommon in partnerships. Do not be blind to another person's deficiencies simply because that person is a friend or family member. Your closeness to each other can intensify your arguments. The success of your business will almost always depend upon your ability to work favorably with your partner.

To avoid any unforeseen misunderstandings in a partnership, it is always best to draw up a written contract. This agreement should clearly state the responsibilities and expectations of each partner. To be safe, it should be drawn up by, or at least approved by, an impartial attorney. An oral agreement and a friendly handshake are insufficient devices upon which to hang your hopes and dreams. The peace of mind you will receive because your contract has been committed to writing will be worth the time, effort, and money it took to draft.

Corporations

A corporation is perhaps the most complicated business structure to form. The individual or individuals who wish to set up the corporation are required to file for a charter in the state where it will operate. The activities that the corporation is allowed to transact are spelled out in that charter. The applicant must first prepare a *certificate of incorporation* with the Secretary of State and pay the appropriate filing fees. The certificate of incorporation lists such items as the names of the parties applying for the incorporation, as well as their home addresses, the business address, the corporation's purpose, what type of activities they will be engaged in, and the kind and amount of authorized stock. Check with state authorities to ascertain how many individuals are required to form a corporation. In many cases, as little as one director is enough to establish a corporation.

A corporation is a separate, legal entity that exists as its own distinct unit. From a legal perspective, it is an artificial person with rights all its own. The corporation exists apart from its shareholders. Outside individuals may purchase stock in the company, but they will usually not affect the existence of the corporation.

A corporation, by the mere fact that it is an artificial entity, is taxed as a separate taxpayer. By virtue of this fact, there exists a possibility of double taxation. Not only will the corporation pay taxes on the money it earns, but also the owners of the company will have to pay taxes on the dividends received. (This possibility can be eliminated if the company elects to become an S corporation.)

There are several advantages to becoming a corporation. Among the most prominent is that the liability of stockholders in a corporation is limited to their capital investments. In addition, a corporation that is profitable has the potential to acquire additional capital through investors or banks. Also, only designated officers or agents can bind the corporation in a contract.

On the negative side, forming a corporation is expensive and must be done according to strict laws. The corporation must follow government regulations and must file its own tax returns and reports. Also, unless the charter is written with a very broad scope, the activities of the corporation could be very limited. Only the directors of a corporation can declare distribution of the profits to shareholders in the form of dividends. Naturally, each shareholder is entitled to the dividend only in proportion to the amount of stock that is owned; but if the directors do not wish to issue dividends, the shareholder is affected by that decision. Moreover, corporations are required to keep very accurate records and to have annual stockholder meetings to satisfy government requirements.

A corporation may elect to become an S corporation if it meets the required qualifications. One of the primary advantages of changing from a regular corporation to an S corporation is that any profits or losses are not taxed directly to the corporation, but are "passed through" to the individual shareholders. Consequently, for example, any loss sustained by the corporation can be used to offset any other income a shareholder might have. By the same token, any profit will raise the income reported on the shareholder's tax return. Keep in mind also that because the business structure is still a corporation, there is limited personal liability. If you have considerable personal assets, it might be a good idea to form an S corporation. However, be aware of the fact that the ". . . corporate veil can be pierced," thereby putting you very much at risk, because you essentially are the corporation.

Forming a corporation does not allow you to operate in any fashion you see fit with impunity. On the contrary, you could be held liable for any acts of fraud or negligence you commit while operating your business. Furthermore, most financial institutions and businesses that extend credit insist that someone from the corporation personally sign the documents covering a debt. If you personally guarantee a loan, then you are liable for that debt even if the corporation files bankruptcy. Therefore, be careful when you sign credit applications and loan documents. However, note that herein lies an interesting dichotomy. If you do not personally guarantee the loan or credit extension, then you will never get the money or the goods. If you do guarantee it, then you risk your personal assets. At that point you have to take a serious look at the situation. The question you have to ask yourself is, "If I cannot pay for this obligation from the corporate funds, will I be able to pay for it out of my personal funds without causing myself undue harm?" If the answer is yes, then sign it. If the answer gives you reason to pause, then reconsider your options.

When you take all these issues into consideration, you really have to analyze whether forming a corporation is truly to your advantage. Given that you may not be as well protected as you might think, you have to wonder whether it is worth the extra expense and complications.

Before you choose a legal structure on which to form your company, consult a competent attorney or business adviser to evaluate your current condition and future plans. The features presented here regarding the formation of a business entity are merely a broadbrush stroke on the canvas of a much more complicated subject. Also, remember to contact the appropriate government agency if you have further questions. Once you choose a business form, you will want to stay with it for a long time. Therefore, pick the correct one from the start, and it will make your life much easier.

Financing your venture

Every year more businesses fail because they are undercapitalized than for any other reason. If you think you can go into business with a couple of months' rent and a few thousand dollars extra, you are in for a shock. The floor covering business is a highly competitive industry, and it is getting more so everyday. If your business is your sole source of income, it is prudent to have at least a sizable amount of cash available for initial start-up expenses plus several months of living expenses set aside. You do not want to use up all your resources within the first few months and find yourself scrambling for a meal. Before you even consider starting your own business, write out a list of all available funds and assets, to ascertain whether you have enough to make it a viable concern. Do not be too anxious to get your business started before you look at what you will need the funds for and what financial options are available to you.

Expenditures

Any new business will have a great many initial capital expenditures. Although most of these expenses are rather costly, they are fortunately only one-time payments. Moreover, many of the initial purchases for such items as sample racks, office equipment, machinery, computers, and inventory actually increase the assets of the company. So, while you may think that you are only spending money, in actuality you are really bolstering the financial stability of the business because all those expenditures are posted to the asset section of your balance sheet. These are some of the many items and services you must consider getting to begin a flooring operation:

- All appropriate federal, state, and local licenses and permits
- Building renovations
- Store display fixtures and racks
- Flooring samples
- Office equipment
- Forklift
- Initial rent and security deposits
- Any professional fees (i.e., legal or accounting)
- Outside and inside display signage
- Promotional and advertising costs
- Security system, if necessary
- Insurance—general liability, worker's compensation, vehicular
- Salaries—for yourself and others
- Taxes
- Utility deposits, telephones, postage, and office supplies
- Business equipment

While this is a rather lengthy list, it does not include everything, and you should add at least 25 percent or more to the dollar total for surprises. As the old saying goes, it's better to be safe than sorry.

Once you have estimated your projected expenses, gauge your anticipated sales volume. This is difficult because a new business has no way to calculate how much sales activity it can generate. That's one reason why purchasing an existing business with a true, verifiable sales record is a good investment. This history is a type of measuring device by which to judge your future sales figures. If you have no way of predicting your sales amounts for the first year with some kind of certainty, you can only hazard a guess.

The time it will take you to generate a profit should also be evaluated. There is a real possibility that for the first several months, little or no cash flow will be produced. If you are in an overly crowded market with too many floor covering stores, it will take some time before you are able to establish a customer base. The longer it takes for your business to become recognized, the greater the chances of failure. If you are operating on a shoestring budget, your capacity to finance it may be limited. Therefore, to arrive at a reasonable amount of money required for you to be successful, you must be brutally honest when projecting potential sales and expense figures.

Sources for obtaining capital

Now that you have taken stock of your financial condition, if your cash on hand is insufficient to cover all your costs and still keep the company afloat for a decent period, here are some possible sources of capital:

- *Second mortgage on your house.* This is very risky because if the business fails, you could lose your home as well.
- *Loan against a life insurance policy.* If your life insurance policy has a cash value, you can borrow against it.
- *Credit or trade union.* If you are affiliated with an organization of this nature, you may be able to secure a loan.
- *Relatives, friends, or business associates.* Mixing business with close personal ties is a very touchy subject. Sometimes it's your most reliable source of capital, yet it could turn into a personal mess if the business fails.
- *Bank loans.* Banks may be very helpful in regard to loans for equipment, machinery, and vehicles, but may be more reluctant to furnish start-up capital. Unless you have impeccable credit and some form of collateral, it may be difficult to get a bank loan. However, if you are going to try this approach, contact a loan officer at your bank and present a well-prepared business and financial plan. The bank will look at your credit history, your business experience, and your ability to repay the loan. Any collateral or cosigners should be clearly indicated. In addition, you should be able to verify that your business plan is sound and that you have the means and the knowledge to make it successful.
- *Small Business Administration* (SBA). The SBA is an independent federal agency that gives advice and guarantees business loans. It does not finance the loan itself. It guarantees up to 85 percent of the loan. It's geared to help minority-owned businesses, but anyone can apply. The SBA also provides free financial counseling services. To qualify for loan assistance from the SBA, you must demonstrate

that you cannot secure financing from any other source. Although you may drown in a sea of bureaucratic red tape, if you qualify for an SBA loan, it could be a great source of revenue. Contact your local SBA office for more details and assistance requirements.

- *Trade credit.* Contact your flooring suppliers to establish a line of credit to help finance your purchases. This could be your greatest source of financial help. It also helps build your credit history. (Also, find out what they can provide in the way of samples and displays—at little or no charge to you.)
- *Issue stock.* If you form a corporation, you may be able to raise capital through a stock issuance.

Accounting and record keeping

The future of your business will largely depend upon your ability to keep accurate records. The importance of a reliable accounting system cannot be overstated. The ability to know how profitable your business is at any given moment can be instrumental in your decision-making process. For example, as you look at a profit-and-loss statement, you can not only see sales figures for that period, but also expenses and how much profit you made. From that information, you will instantly realize whether to increase your sales volume or trim certain expenses.

The exact type of system for you to use should be approved by your accountant. Whether you choose a single- or double-entry bookkeeping system will be determined by the size of your operation and the legal structure of your company. A corporation, for instance, is required to use the double-entry system, while a sole proprietorship does not have the same restriction. Also, whether you wish to do your books by hand or use a software program developed for a computer is a major consideration. Posting all your transactions by hand with pencil and paper is a long, laborious process. Modern computer programs are generally much faster and more efficient. There are also programs on the market written specifically for the floor covering trade. Consult the flooring industry trade journals for information regarding these products. However, be careful before you purchase these items, because they are often rather expensive and may not give you any more assistance than you would normally get from an over-the-counter bookkeeping program.

Familiarizing yourself with fundamental accounting principles will help you determine whether a certain accounting system is right for you. Regardless of whether you choose to use the hand method or the computer method, the basic principles are still the same. You have to make entries into journals and ledgers in order to calculate the profit-and-loss statement and prepare a balance sheet. (For the purposes of this discussion, assume you will be using a double-entry bookkeeping system. Bear in mind that this information is just a brief introduction to a very complicated subject. Consult a professional accountant for more information.)

The fundamental types of records include five journals and a ledger system:

- Cash disbursements journal (Fig. 8-2)
- Sales journal (Fig. 8-3)
- Cash receipts journal (Fig. 8-4)
- Purchase journal (Fig. 8-5)
- General journal (Fig. 8-6)
- Accounts ledger (Fig. 8-7)

Individual transactions such as sales, purchases, and cash income are reported in the journals daily. At the end of the month, each journal is tabulated, and the totals are posted to the respective ledger accounts. The amounts taken from each separate ledger account are then transferred to the corresponding account listing on the profit-and-loss statement or the balance sheet to create the monthly financial reports. (Most computer programs post each transaction to the ledger account as soon as it becomes a journal entry.)

In the double-entry system, for every transaction there are always two separate entries: a debit and a credit. These entries are always for an equal amount. Therefore, an accounting maxim is: the debits always equal the credits. If they do not, there is an error. You cannot proceed until you have discovered and rectified the error. The precision of this system makes it a model of certainty. It either balances (debits equal credits) or does not. It is an absolute. There is no room for hypothesis or conjecture. It is either right or wrong. Finding an error can be an interesting, yet time-consuming affair. However, it's always fascinating when you do find the error. Quite often it is a simple mental mistake, such as a *transposition* (writing 92 instead of 29).

A brief look at the journals and ledgers reveals their functions:

- *Cash disbursements journal.* This journal records all payments made by the company.
- *Cash receipts journal.* It records payments made to your company by your clients or by any outside source.
- *Sales journal.* This records sales on account.

- *Purchase journal*. It records purchases made by your company on credit.
- *General journal*. This journal records any transaction that is not normally appropriate for any of the other journals.
- *Accounts ledger*. The ledger is a series of separate, one-page accounts that divide the transactions into identifiable record-keeping units. For example, the ledger consists of single-page accounts such as the following: Cash account, Inventory account, Accounts Receivable account, Equipment account, Proprietorship account, Accounts Payable account, Loans Payable account, and all the expense accounts such as Rent Expense account, Advertising account, Insurance Expense account, and so on.

The ledger is divided into five types of accounts; asset accounts, liability accounts, proprietorship accounts, expense accounts, and earning accounts. Each has a prevailing balance that is a debit or a credit. (A debit simply means the left side of an entry, while a credit means the right side of an entry.) For example, expense accounts generally have a debit balance while sales accounts have a credit balance. Since for every debit there must be a credit, by allocating certain accounts with prevailing types of balances, it helps make the debits and credits equal one another. Sir Isaac Newton's third law of motion states that for every action, there must be an equal and opposite reaction. This principle illustrates perfectly how the rules of accounting work: What goes out pays for something that comes in—and for an equal amount.

To illustrate this point, assume you are about to enter into a transaction whereby you are to sell Mr. Hamilton a roomful of carpet. Since you want to make a profit, several transactions must occur and be posted to the appropriate ledger accounts (in equal amounts) to complete the cycle. If you agree to buy the carpet from your supplier at $1000, you pay $1000 in cash to receive $1000 in merchandise. Then when you sell the carpet to Hamilton for $1350, you will receive $1350 in cash from him for a sale that amounts to $1350. Hence, if you post this transaction to the General Journal, you record it as follows:

Debit—Merchandise inventory (received) $1000
Credit—Cash (paid out) $1000

Then, upon the sale of the goods, you record

Debit—Cash (received) $1350
Debit—Sales (merchandise going out) $1350

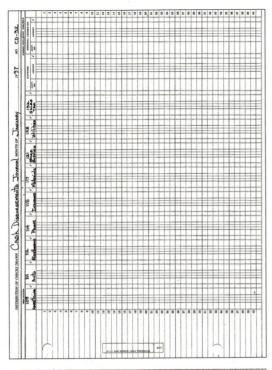

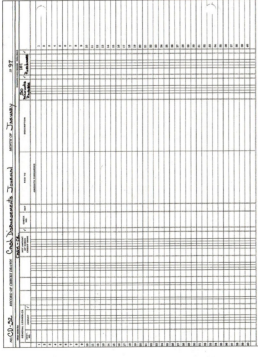

8-2 *Cash disbursements journal, two parts.*

237

No. CD-32   RECORD OF CHECKS DRAWN   Cash Disbursements Journal   MONTH OF JANUARY   19 97

STANDARD CHECK RECORD   FORM 264-G

GENERAL LEDGER-CR.		CASH - CR.						AMOUNTS FORWARDED		50 Accounts Payable		101 Purchases	
ACCT. NO.	AMOUNT	✓	NET AMOUNT OF CHECK CREDIT BANK	✓	CHECK NO.	DAY	PAID TO	DESCRIPTION	✓	✓		✓	
1											1		
2											2		
3											3		
4											4		
5											5		
6											6		
7											7		
8											8		
9											9		
10											10		
11											11		
12											12		
13											13		
14											14		
15											15		
16											16		
17											17		
18											18		
19											19		
20											20		
21											21		
22											22		
23											23		
24											24		
25											25		
26											26		
27											27		
28											28		

8-2 *Continued.*

238

DISTRIBUTION OF CHECKS DRAWN

Cash Disbursements Journal — MONTH OF JANUARY 1997 — NO. CD-32

STANDARD CHECK RECORD FORM 264-G

109 Advertising	111 Auto	112 Miscellaneous	114 Phone	115 Insurance	117 Materials	121 Store Supplies	128 Utilities	152 Sales Tax	EXPENSE — ACCT. NO.	EXPENSE — AMOUNT	GENERAL LEDGER-DR. — ACCT. NO.	GENERAL LEDGER-DR. — AMOUNT
1												
2												
3												
4												
5												
6												
7												
8												
9												
10												
11												
12												
13												
14												
15												
16												
17												
18												
19												
20												
21												
22												
23												
24												
25												
26												
27												
28												

USE STANDARD POST BINDER BPS 1114

8-2 Continued.

239

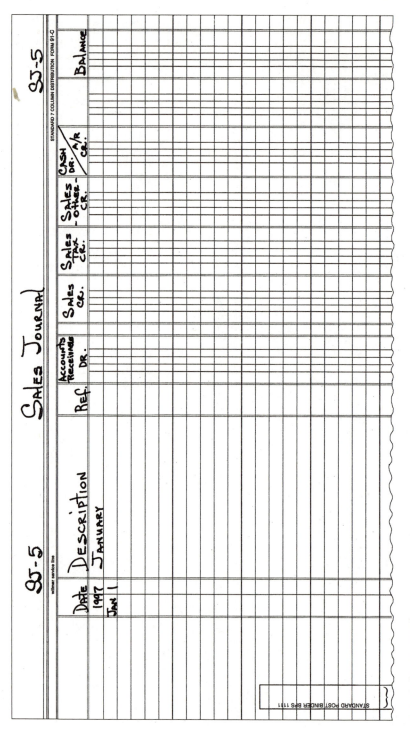

8-3 *Sales journal.*

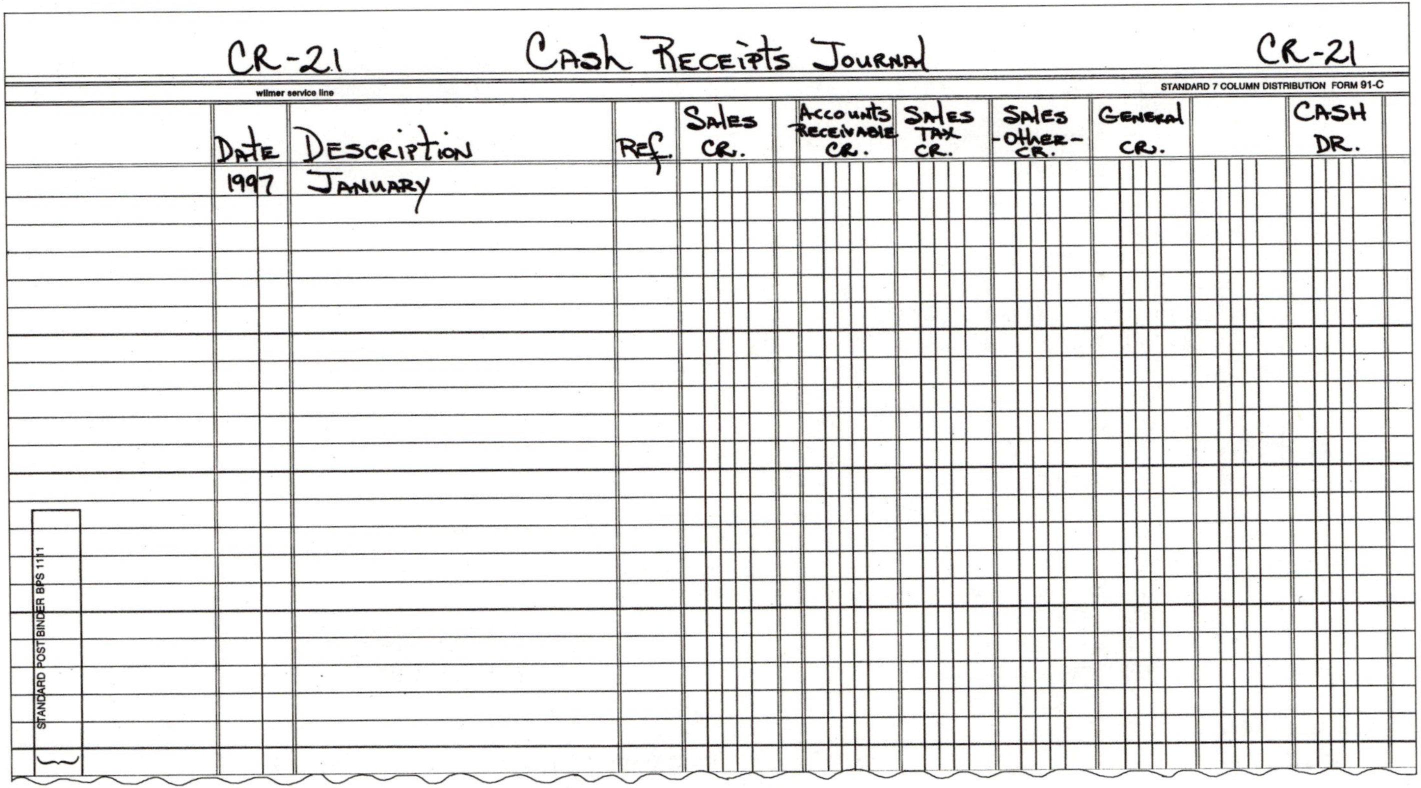

Cash Receipts Journal

CR-21

wilmer service line

STANDARD 7 COLUMN DISTRIBUTION FORM 91-C

Date	Description	Ref.	Sales Cr.	Accounts Receivable Cr.	Sales Tax Cr.	Sales - Other - Cr.	General Cr.	Cash Dr.
1997	January							

STANDARD POST BINDER BPS 1111

8-4 *Cash receipts journal.*

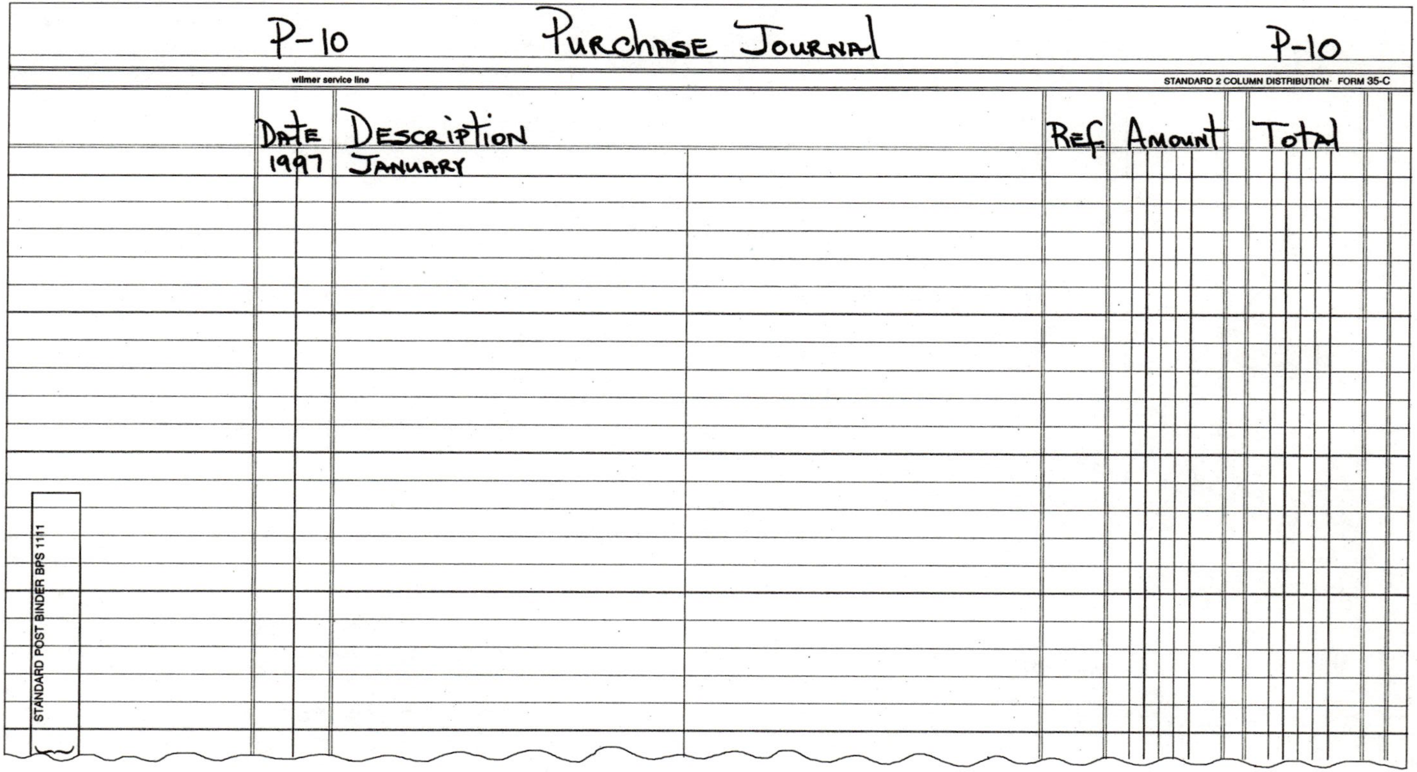

P-lo   Purchase Journal   P-10

STANDARD 2 COLUMN DISTRIBUTION FORM 35-C

wilmer service line

Date	Description	Ref	Amount	Total
1997	January			

STANDARD POST BINDER BPS 1111

8-5 *Purchase journal.*

GJ-8   General Journal   GJ-8

Date	Description	Ref.	Debit	Credit
1997	January			
Jan 1				

8-6 General journal.

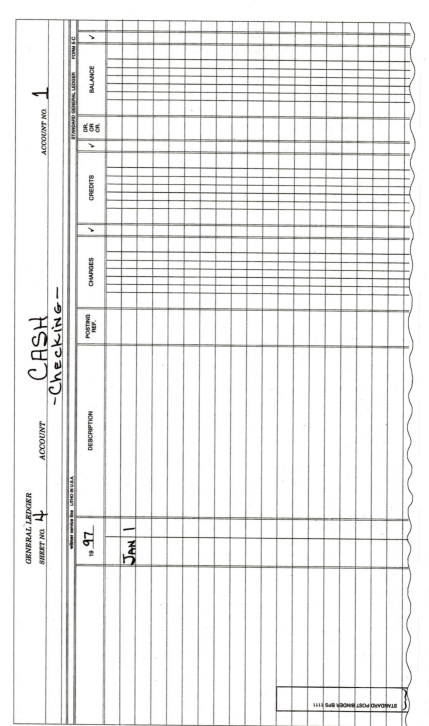

8-7 *Ledger accounts, six parts.*

244

GENERAL LEDGER

SHEET NO. 7   ACCOUNT   Inventory   ACCOUNT NO. 6

wilmer service line   LITHO IN U.S.A.   STANDARD GENERAL LEDGER   FORM 5-C

19 **97**	DESCRIPTION	POSTING REF.	CHARGES	✓	CREDITS	✓	DR. OR CR.	BALANCE	✓
Jan 1									

STANDARD POST BINDER BPS 1111

8-7 *Continued.*

245

GENERAL LEDGER

SHEET NO. 12   ACCOUNT   Accounts Payable — Merchandise —   ACCOUNT NO. 50

wilmer service line   LITHO IN U.S.A.

STANDARD GENERAL LEDGER   FORM 5-C

19 97	DESCRIPTION	POSTING REF.	CHARGES	✓	CREDITS	✓	DR. OR CR.	BALANCE	✓
Jan 1									

STANDARD POST BINDER BPS 1111

8-7 Continued.

246

GENERAL LEDGER

SHEET NO. 3   ACCOUNT   RETAINED EARNINGS   ACCOUNT NO. 81

STANDARD GENERAL LEDGER   FORM 5-C

wilmer service line   LITHO IN U.S.A.

19 97	DESCRIPTION	POSTING REF.	CHARGES	✓	CREDITS	✓	DR. OR CR.	✓	BALANCE	✓
JAN 1										

STANDARD POST BINDER BPS 1111

247

8-7 *Continued.*

GENERAL LEDGER

SHEET NO. 18 ACCOUNT Purchases ACCOUNT NO. 101

Wilmer service line LITHO IN U.S.A.

STANDARD GENERAL LEDGER FORM 5-C

19 97	DESCRIPTION	POSTING REF.	CHARGES	✓	CREDITS	✓	DR. OR CR.	BALANCE	✓
Jan 1									

STANDARD POST BINDER BPS 1111

248

8-7 Continued.

GENERAL LEDGER

SHEET NO. 5   ACCOUNT   Advertising

EXPENSE

ACCOUNT NO. 109

wilmer service line   LITHO IN U.S.A.

19 97	DESCRIPTION	POSTING REF.	CHARGES	✓	CREDITS	✓	DR. OR CR.	BALANCE	✓
Jan 1									

STANDARD GENERAL LEDGER   FORM 5-C

STANDARD POST BINDER BPS 1111

249

8-7 Continued.

Now that this transaction is completed, you can see that although you made a $350 profit, each segment of the transaction recorded equal dollar values. The $350 profit will not be readily apparent until the financial reports are prepared. At that time $350 will appear as the difference between the cash paid for the merchandise and the cash received for the merchandise; or since there are always two ways to look at a transaction, it could be the difference between the merchandise cost and the sales price. Either way, it's always one debit versus one credit: In this analysis it's a greater amount versus a lesser amount (i.e., a $1000 debit versus a $1350 credit that will show where the profit comes from.) Only you are going from one transaction to the next—the debit from one transaction and the credit from the other transaction.

Financial statements

Once all the individual transactions have been posted to the appropriate journal and the totals of each section are posted to the corresponding ledger accounts, it's time to prepare the financial statements. These statements, which include a profit-and-loss statement along with a balance sheet, should be prepared monthly and then once at the close of the year to give an annual picture of the company's financial situation. From these reports you can see where you are at present, where you were in the past, and where you should be in the future. They provide accurate and comprehensive information on all financial activities of the company.

The balance sheet is derived from the assets, liabilities, and proprietorship accounts from the ledger accounts. The profit-and-loss (P + L) statement is made up of the income and expense accounts. The correct sequence is to prepare the P + L statement first and then the balance sheet. The reason is that the net profit (for a given period) deduced in the P + L statement will be inserted into the proprietary section of the balance sheet. Without the net profit for that period, the balance sheet would not balance.

A balance sheet reports a company's financial condition at a particular date (say, June 30) whereas a profit and loss statement shows the activity of the business from one time to another (e.g., from January 1 to June 30). The balance sheet lists the assets (debits) and liabilities (credits) of a company. The difference between the assets and the liabilities is the net worth (or proprietorship) of the company. The basic equation of a balance sheet is Assets = Liabilities plus Proprietorship ($A = L + P$).

The object of the profit-and-loss statement is to find the net profit. This is achieved by taking the total sales figure and deducting the cost of goods sold. This tells you the gross profit. From the gross profit you will further deduct any expenses. This will determine what your net profit truly is. As stated earlier, this figure is transferred to the proprietorship account in the balance sheet to help balance that report.

A rudimentary understanding of basic accounting principles will be of inestimable value as you travel the road of entrepreneurship. Take the time to further study this fascinating field, and you will reap valuable rewards. Not only will it help you to make vital business decisions, but also it will protect you against unscrupulous individuals who have injurious intent. Far too often, an owner of a company lacking adequate accounting skills, has been hoodwinked into financial disaster by dishonest employees. A knowledgeable bookkeeper can "doctor" your books to such an extent that you won't know what happened until it's all over! However, if you know what to look for when you examine your books, perhaps you can prevent such unpleasant situations; don't just look at the bottom line and think everything is fine. For example, the person who is preparing the reports, and writing the checks, could be the very person who is embezzling funds from you by inflating the expenses through a "dummy" expense account. It wouldn't be the first time it happened to some unsuspecting, inexperienced business owner. Therefore, learn what you need to know about accounting that can help you spot potential problems before they get out of hand. Remember, it's your money, so protect it as best you can. You work hard to make a profit; don't give it away simply because you don't want to take the time to study accounting. Furthermore, make sure you sign all the checks yourself, before they are sent out. That is the best form of control you could ever have. That way, no money leaves your company without your knowing where it's going.

Setting up shop

Once you have all the preliminary details worked out, it's time to set up shop. Much will need to be done to get the showroom ready and to develop a viable organization. To begin, contact the major suppliers you wish to do business with and ensure that you will have no trouble obtaining their goods. Many times, particularly in small communities, certain manufacturers have developed relationships with dealers in that area that they do not care to jeopardize. If you count on doing business with that manufacturer, only to find out that it does not wish to sell to your company, you could have a major problem. We are seeing this more and more today as manufacturers and dealers are forming exclusive partnerships with one another. These types of relationships could prevent you from purchasing certain products. Therefore, make exploratory contacts with all your intended suppliers to eliminate this possibility.

If you are planning on becoming a full-service flooring operation, the suppliers should be able to provide the following products: carpeting, sheet vinyl and other resilient floor coverings, ceramic tile, hardwood products, and any specialty flooring materials you think could boost your sales figures. As your business expands, you may wish to add other items such as window coverings or other products that complement the merchandise selections available to your customers. Consult trade directories found in most local libraries for the phone numbers and addresses of the manufacturers you are interested in. Once you have successfully contacted them and they have agreed to sell you their products, you need to obtain two things from them: credit and samples.

Credit

Trade credit, also referred to simply as *credit,* is essentially a delayed-payment plan established by suppliers that allows a purchaser a given time within which to pay for the goods. It is, in effect, a short-term loan. For a business that has limited funds, taking advantage of trade credit is an excellent way to retain working capital for as long as possible. Flooring retailers that are unable to secure a line of credit with a supplier are forced to pay for their purchases on a *cash-before-delivery* (CBD) or a *cash-on-delivery* (COD) basis. As the names imply, when a company is on a CBD basis, the supplier requires payment in full *before* the materials are shipped. When a company is on a COD basis, the goods must be paid for when the materials arrive. It is often required that COD shipments be paid for by cash or a cashier's check. Company checks are often not accepted. The reason is that there may be insufficient funds in the purchaser's checking account. The supplier would not know this for several days, and then it might be much harder to extract payment from the purchaser who has possession of the goods.

Trade credit in the flooring industry typically provides *terms.* *Credit terms* refer to a payment incentive program devised to encourage purchasers to make timely remittals. Suppliers give the purchaser a small cash discount if the merchandise is paid for within a specified period. For example, terms common among many suppliers are 2/10, net/30. When terms of this nature appear on an invoice for merchandise received, it means that the purchaser is entitled to a 2 percent cash discount (also known as a *discount*) if the invoice is paid within 10 days of the date on the invoice. The cash discount is merely deducted from the amount owed. If the invoice is paid within 30 days, no cash discount is allowed, payment must be made in full, but the account is not considered delinquent. However, if payment is received by the supplier more than 30 days after the date of the invoice, the account is considered delinquent. If payments are repeatedly delinquent, this could affect a company's credit rating and subsequently lead to complete loss of credit privileges. If this happens, the company will be placed on a CBD or a COD basis until it has demonstrated sufficient care in handling its payments.

When a company is put on a cash-before- or a cash-on-delivery basis, it could create some awkward situations. The most unpleasant aspect about this predicament is that unless you can convince your customer to pay you the entire amount due for the flooring up-front, you will have to pay for it out of your own funds. In this case, you are financing the purchase until it is installed and paid for. If, for any reason, there is any delay in payment to you from your customer, it could hinder your ability to operate freely because your financial resources are tied up. Your customers, however, will probably be very reluctant to pay the entire amount for the merchandise until it has been installed. Since most flooring products are installed by the store that sold them to the consumer, it is not common for the store to be paid in full until the customer is satisfied with the workmanship. Inferior merchandise, poor installation technique, and inadequate customer service are only a few of the problems that could delay final payment.

One main drawback of being on a COD basis is that you have to take time to either gather the hard cash or go to the bank to get a cashier's check, *and* you have to be at the location with the money in hand when the driver arrives to deliver the merchandise. This could create logistical problems or cause the driver to return to the delivery terminal with your merchandise still on the truck. If you or someone from your company does not pay for the goods upon delivery, the driver cannot leave the merchandise with you. And sure enough, the merchandise you need delivered on a specific day, because you are scheduled to have it installed that afternoon, is going to be the merchandise that winds up back in the terminal— undelivered. This will undoubtedly cause numerous problems that will be difficult to rectify. Therefore, whenever possible, establish good credit and keep it.

Paying invoices on time and taking the credit terms can affect your overall profitability in several ways. First, if you "take" your discount of 2 percent, you add that much more profit to the overall job; it's 2 percent you didn't calculate in your original proposal. At the end of the year, that could prove to be a sizable amount. Second, if you are bidding on a job that is going to be bid on by several other contractors who quote very tight prices, you might want to figure your cost less that 2 percent in the hope of getting the job. However, this could be a very risky approach. Not only must you be sure you can pay for the merchandise on time out of your own funds in case the buyer's payment is delayed, but also if there happens to be some unforeseen administrative error and the bill is not paid on time, you will have just lost 2 percent of what could have been a significant portion of your profit. On a big commercial job that could be a lot of money. Consequently, it's often best to just figure that 2 percent as a bonus and add it to your bottom line. If, for some reason, you are unable to pay the invoice within 10 days, you're really not taking away from the profit amount you originally calculated for the job. It's just unfortunate you were not able to pick up the extra percentage points. The most important thing, however, is to pay off the invoice within 30 days to keep your credit rating intact.

Samples

To offer anything for sale, you must have adequate samples. Flooring companies will provide these samples for you so that you can begin selling their products. What you will have to pay for these samples is negotiated by you and the manufacturers' territory representative, better known as the *sales rep*, or simply, the *rep*. The sales representative for each individual manufacturer is your conduit to the supplier. They can get you samples, display racks, and special pricing and can assist you with solving problems regarding their merchandise. A good working relationship with your sales representative can be very beneficial to the success of your company. Most representatives are congenial, hard-working individuals who want to help you make a profit, which in turn helps them by increasing their sales figures and commissions. (Some representatives are on salary, but their increased income depends on high sales figures.)

The kinds of samples you need will depend upon the size of your showroom, the products you wish to carry, and the funds budgeted for initial sampling expenses. Remember, you can always add more samples as time goes on. So don't overdo it before you are making a profit and can justify more samples. Sometimes, less is more—providing you offer a broad array of merchandise that will satisfy the tastes of a vast majority of your intended customer base.

The store of the future will be spacious, yet compact. It will offer nearly every conceivable form of flooring material in an uncluttered, contemporary-looking showroom. Displays will be arranged in such a fashion that customers will be able to browse through the show-room without the need for overbearing salespeople thrusting sample after sample at them. Display racks containing large enough samples to give the viewer a realistic representation of how the flooring will look in their home will be the norm.

How flooring dealers merchandise their wares will be a pivotal factor in their ability to garner market share in an increasingly com-petitive industry. As consumers become more sophisticated in their buying habits, how a dealer satisfies their need to properly visualize the product in their home may very well be a determining factor in their decision-making process. Which dealer gets the sale may de-pend upon who has the best merchandising program. On the other hand, no matter how great your showroom looks or how many large samples you can heap upon your customer, the bottom line is still *price*. How you price your products can make or break your business.

Pricing

The price you set for your products will largely depend upon your marketing strategies and the type of client to whom you wish to ap-peal. If you are interested in attracting the *low-end* market, which in-cludes property management, builders, and lower-income clients, then your pricing markups will have to be lean. In this market, the goal is to generate enough volume to compensate for the low markup. However, if the markups are too low and the volume drops (or disappears), your cost of doing business will soon outweigh your profits. The low-end market, therefore, is a volume-driven market. Securing a contract with a large property management firm is a way of establishing a base in this market.

Conversely, the *high-end* market is a more selective segment where volume will be lower, but markups can be greater. Fewer jobs, at a greater profit margin, can often equal or outweigh those of the low-price, high-volume variety. Less work for more money is cer-tainly an entertaining thought, but not every dealer can achieve that result. The high-end market requires a firm and loyal customer base of wealthy clients. Few flooring retailers have that base. Further-more, the high-end market requires exceptional customer service and attention to detail. People will not pay overinflated prices for long if the service and installation of the products are inferior. You might get away with it on a few jobs, but in the end it will catch up with you. Therefore, if you are going to ask a high price, back it up with value and excellence.

The optimum approach then is to attempt to attract both the low- and high-end markets as well as the *middle market*. By appealing to all customer types, you can increase your market share at will. Filling your showroom with low-, middle-, and high-end products will allow you to market your merchandise to the greatest number of people.

No matter which market type you finally target, your prices must be in line with what the customer regards as *real value,* as opposed to *perceived value*. Real value provides the buyer with a product that is not only priced well, but also one that has quality and performance characteristics that are desirable. Perceived value, on the other hand, gives the customer a false sense of assurance because it misleads those individuals who are unfamiliar with a product's features into believing that a particular flooring item is better than it really is. For example, a carpet that is manufactured using an inferior quality fiber, but with a heavy "hand," may appear to have a greater perceived value than a carpet that is produced with an advanced-generation fiber having a lesser "hand" but with a higher yarn twist level. In reality, the latter carpet is a much better product, but to the uninitiated it may not appear so. Therefore, it is imperative that you provide your clients with products that have real value based on style, quality, performance, and price. However, this is not to say that perceived value is a totally negative term. It does have its use in the marketplace because not everyone is looking for long-wearing, high-quality products. A homeowner, for instance, who wants to spruce up a home in order to sell it may want a carpet with a high perceived value. A carpet that looks more expensive than its cost may be just fine in this case. Here, the concept of perceived value takes on a positive connotation. Consequently, it is important for you to consider all aspects of value interpretations when choosing products you wish to offer for sale to your customers, and when formulating the framework for your pricing strategy.

While real and perceived values may be intangible commodities, to price your products correctly, you must include such tangible items as freight charges, merchandise costs, labor costs, and sales tax, as well as your overhead costs like rent, insurance, etc. Prices, therefore, must be based on quality, style, and service. What you offer—in contrast to what your competitors offer—will provide the basis for your pricing structure.

What's important, however, is to keep your competitors' prices in mind, but to use your own criteria to set your prices. If you are pricing your products from a defensive position, that price may not cover all your costs and still provide you with a reasonable profit. But if you do not price your goods competitively, you run the risk of losing too many jobs and may soon find yourself in tremendous trouble. So devise a formula that works for you and stick with it, if it proves to be productive. As they say, nothing succeeds like success!

Marketing and advertising

How you market and advertise your business will ultimately lead to its success or failure. Marketing and advertising are two entirely different concepts. *Marketing* is the process of studying and evaluating the market for your products or services, whereas *advertising* is the physical act of communicating with potential clients. Marketing involves research into such aspects of the marketplace as customer buying trends, household income demographics, determination of products that buyers currently are interested in, tabulating advertising results, and much more. Marketing is an attempt to identify a specific segment of your community, called a *target market,* that is most likely to purchase what you sell. Once you have identified that target market, cover it with advertising. Your market research should help determine what form of advertising will be the most effective approach—Yellow Pages, newspaper, radio, television, direct mail, telemarketing (telephone sales), coupons. Regardless of the size of your business, an understanding and use of both marketing and advertising techniques is essential.

Marketing

The focus of your market analysis is to determine what the buying public wants. Once you have figured that out, you can implement programs that will bring people to your door and make them buy from you. That's what it's all about. You want to induce sales. The whole reason for the existence of your establishment is to generate sales. A going concern already has an established customer base; for a start-up company, fresh research is necessary. In either case, the main idea is to expand the customer base; one company is starting from zero, and the other company is starting from a certain point in regards to sales figures. (For the purposes of this discussion, assume that the company in question has a small, established customer base that needs to be expanded.)

When you do a market analysis, it is necessary to evaluate (1) the current products offered for sale and (2) any new products that can be introduced to generate more interest and to (3) identify your current customers and (4) your prospective customers. If you are going to sell just the same old products to your current list of customers, your business is going to stagnate and will soon flounder. If you wish to improve your market share, you must find new products to sell to your existing customers and open new markets where you can sell both the old and the new products. By doing this, you increase your prospects exponentially. Therefore, once you have all the latest products at your disposal, your main thrust should be to open new markets.

Before you begin to randomly search for additional clientele, study your current customer profile. Your best likelihood of attracting new customers is to target people with characteristics similar to those of your present buyers. Once you have determined this customer profile, the purpose of your market research is to identify who these people are, where they are, and by what means they can be reached. You may need to consult census reports or trade journals or contract with a research consulting firm. Your objective is to know as much about your potential customers as possible, so you can appeal to and meet their needs. The format of that appeal is advertising.

Advertising

Advertising and marketing go hand in hand. What you learn in your market research should be implemented in your advertising. If, for example, a company identified its target market as suburban couples in their middle-forties, then it might choose to catch their attention in radio advertisements by playing background music of songs prominent when those couples were teenagers. Likewise, if your target market is middle-income families of Hispanic descent, then music with a Latin flavor might be more appropriate. The point is, no matter what your research tells you, be sure to utilize that knowledge in any way you can. Remember, too, that market research does not have to be some elaborate project costing thousands of dollars. It could be as simple as studying your prior customer records and coming up with a composite profile. That profile should indicate *who* your customers are, *what* they purchased, *where* they live, *when* they bought, and *how* they came to hear about your company. With that information you can develop an effective advertising campaign.

Successful advertising should draw the buying public's attention to your business, and create an interest and desire in your product that will encourage buyers to take action. First-rate advertising projects a positive image of your company which induces people to want to do business with you.

Whichever advertising medium you choose, whether it be radio, newspaper, television, or the Yellow Pages, keep in mind that each one should be used to achieve particular results. The object is to reach the greatest number of people in your target market at the lowest cost. When pondering what medium to choose, consider the size of the medium's audience, its cost, and the quickness of the results you can expect.

There are two basic types of advertising: in one, a specific offer is made to elicit immediate action; in the other, institutional advertising is designed to promote name recognition. In ads that are placed to create an immediate action, you might want to make the following offer:

> Thick, plush carpet that was $21.00 per yard installed, is now on sale for 3 days only at $10.99 per yard installed with high-quality padding.

In this type of ad, audience members realize that in order to take advantage of this seemingly stupendous offer, they must react quickly and visit your store before the 3 days are up; otherwise, they will have to pay the higher price. In an institutional ad, you might say something like:

> Pete's Carpets—We Sell for Less.

This kind of ad merely gets your name in the public's psyche in the hope that when (and if) they get around to thinking about buying new carpet, they'll remember your ad or logo.

When you are considering both advertising approaches, a combination of both immediate response and institutional advertising could be an effective strategy. By gaining the public's attention with constant reminders of your existence through credible institutional ads, then periodically offering specific response inducements such as a President's Day sale, you could boost your sales figures. Potential customers tend to frequent businesses they are familiar with, even if they have never set foot in your store. That's how powerful a good advertising campaign can be. A business should constantly try to reinvent itself in the public's eye. Introducing itself to new prospects and reintroducing itself to previous clients will help stimulate a constant flow of traffic. It's important to jog people's memory and remind them of your existence.

The key element of a successful advertising strategy is *consistency*. Satisfactory results are unlikely with inconsistent advertising programs. Even if you cannot afford to have full-page, double-truck ads, consistency is still essential. With so much emphasis being placed on high-technology, computer-generated artwork, and massive display ads by major department stores, the viewing public has become desensitized to most ads and sometimes turns the page without stopping to read them. They often leaf right past those pages in the newspaper, only to have their attention drawn to a more cleverly designed, smaller ad. This ad may have a greater impact on their subconscious than the much larger ads do. To design a well-constructed ad, it is necessary to figure out what motivates your customers and what process they use to decide to make a major purchase such as flooring.

When people buy a certain product, they are really buying the benefit they derive from that product. In carpeting, it may be the luxurious feel on their bare feet as they walk across the room. With ceramic tile, it may be a pleasing pattern coupled with the knowledge that it will last for many years. By emphasizing specific benefits and qualities in your advertising, you can tap into the real reason why people buy things—the product's beneficial features.

Another form of advertising is *sales promotion*. Sales promotions can include contests, giveaways, drawings, coupons, displays, games, and the like. Getting the prospective buyer's attention through unique and unusual happenings is always an appealing inducement to get people "in the door." Once they are in, it's up to you to captivate them with your selling skills. Having a clown hand out balloons in front of your store on a Saturday afternoon is a good example of an attention-getting sales promotion tool. It is also an excellent illustration of the difference between advertising and sales promotion. Sales promotion often calls upon your creative ability to find ways to generate interest in your company which are outside the traditional avenues of advertising such as broadcast media and print. In many respects, sales promotions can be more fun and can provide pleasant relief from mundane advertising programs.

Yet another form of advertising that is not really physical advertising is *public relations*. Public relations comprise a company's efforts to create a favorable image in the community at large. This is done by news releases, established relationships with the local populace, and maintenance of customer satisfaction.

If a company can get a local newspaper or radio station to publish a news release about something done by, or in, the floor covering business itself, that is perhaps one of the best forms of advertising. Not only is it free, but also it provides a high degree of respectability and legitimacy. An example of this type of public relations press release might be: "Pete's Carpet Company installs new carpet in the Pre-School Day Care Center after providing the 30 children and teachers with a lesson on what it's like to be a carpet layer. Each child got to observe Pete as he showed how he measures, cuts, and joins, the carpet." The negotiations for the purchase and installation might have included some incentive pricing to the purchaser, with the understanding that Pete would get some public relations coverage from the project. When a company offers to a community some of its resources that improve the quality of life within its boundaries, like donating time and products for good causes, this can enhance the company's image considerably.

Perhaps the most critical part of public relations is the internal handling of customer satisfaction. Being honest with your clients and being courteous to them through times of difficulty are critically important to the survival of any business. By taking care of complaints promptly and putting yourself in your customer's shoes, you may get countless referrals from that one happy customer. On the other hand, an unhappy customer most likely will tell many more people than a happy customer because at that point the unhappy customer has a story to tell and will tell it to anyone who will listen. It's like the evening television news; it always seems more interesting to hear about one misfortune than about the thousands of good things that were done in the world that day. Treat all people politely and with respect, so they may return the favor by speaking kindly of you.

Sales

No matter what title you may wish to bestow upon yourself, be it corporate president, owner, or CEO, the fact remains that you are, above all, a salesperson. Regardless of how large your sales staff may be, you are still the lead salesperson. It is your responsibility to take charge of the sales force in one way or another. You have to motivate others as well as yourself. To do so, you must understand some simple, basic concepts of successful selling practices.

To begin, understand that sales is your function. Therefore, you need to get excited about it. Just as a football running back's goal is to score touchdowns and a basketball player's main purpose is to score baskets, your objective is to make sales. Without sales, there is no company. If you can get just as excited about closing a sale as a football player is about scoring touchdowns, you've made the first step in becoming a proficient salesperson.

Selling floor coverings has both similarities to and differences from selling other products. Persons skilled in the art of selling can sell any product, once they have become familiar with that product's form and function. The techniques of selling remain the same, only the merchandise (or service) changes.

A professional salesperson knows that a good first impression is critical to an auspicious start. The cliché "You never get a second chance at a first impression" certainly applies in sales. Furthermore, sales personnel are often a customer's initial contact with your firm. If that first impression is a bad one, some major hills may have to be climbed to regain that customer's confidence. Therefore, sales representatives should be well groomed, dress neatly, and have good personal hygiene. Facial expressions, body language, and vocal inflections should reflect a calm, assured attitude. Their behavior should be courteous and respectful throughout the entire sales process. The image they present will be the image customers take with them when they think of your company. All customers who walk into your showroom should be greeted with a pleasant "Hello" or "Good morning," to let them know they are welcome in your establishment. There is nothing worse for a potential customer than to walk into a place of business and be ignored. If you are busy, just stop what you are doing for a moment and acknowledge the person's presence. If what you are doing cannot wait, say hello and indicate that you will be with the person momentarily. Never conduct personal conversations or attend to personal business while a client is waiting. It is rude. Put yourself in the client's place and behave toward them as you would like to be treated when you walk into a store.

What is really taking place in a sales transaction is that the customer has a need and it is the salesperson's responsibility to satisfy that need. However, that need takes on many forms when it comes to flooring. For the purposes of this discussion, assume that the buyer intends to purchase an entire house of carpeting. Given this fact, we immediately know the following:

- This will most likely be a major purchase for that person. Therefore, careful deliberations will take place within the family before a decision is reached. These deliberations will take place over an extended time, so an order usually is not placed on the first visit to the store. Consequently a "soft sell" is usually more appropriate in the flooring industry than a "hard sell." (A *soft sell* refers to giving a person ample time to make a buying decision, and it is a *hard sell* when the salesperson "hammers" someone through intimidation and fear into purchasing an item before he or she really wants to place an order.)

- Some of the many needs that must be met are: price, style, color, texture, quality, value, and proper installation.
- The clients will need to see the samples in their own house to verify color coordination with their other furnishings and surroundings.
- They most likely will compare price and features with other flooring dealers.
- How you present yourself in a professional manner to that customer will either help or hurt your cause.
- The customer will require your expert knowledge to help them become an informed buyer.
- The house will need to be professionally measured to verify room dimensions.
- The salesperson must focus on the benefits this product will provide to the buyer, i.e., cleanability, wear, and comfort.

As you can see from these points, buying floor covering is quite often a long, drawn-out process. Care and consideration must be given to the client to help "guide" her or him into making the right decision. Through training and practice, a salesperson becomes skilled at understanding the psychological processes involved in decision making. Good communication is essential to make the transaction a pleasant one.

Once the home has been measured, a carpet selection has been made, and the price has been quoted to the client, it's time to close the sale. It sounds as if it should be a simple matter, but you would be surprised how often a salesperson fails to "ask" for the order. This is the single most important moment in the whole selling process. When you have done all you can and you have carefully overcome any possible customer objections, you have to say the words that will ask for the order. However, you want to phrase it so as to avoid getting a "no" response. Statements such as "Which color would you prefer to order?" or "Would you like to order the heavyweight carpet or the lighterweight version for your home?" or "When would you like us to install this carpet for you?" will do just fine. Begin at once writing the order, after there is an understanding that the sale has been consummated. Then complete the sale by obtaining the customer's signature and taking a deposit on the order. Once these final two aspects of the sale have been fulfilled, it is less likely the customer will back out of the deal. Occasionally, buyers have second thoughts, but they infrequently do so once they have signed their names to a document and given a deposit to the salesperson. Remember, too, that certain elements must exist to have a valid contract.

For a contract to be a true, legal document, five elements must be present: offer, acceptance, written instrument, consideration, and competent parties. The *offer* is the product and services you are offering for sale at a certain price. The *acceptance* of the offer occurs when the customer signs a *written instrument* (contract) prepared by you, itemizing what you are selling. The *consideration* is the monetary deposit you receive from the customer. A *competent party* to an agreement must be of legal age and mentally competent to enter into the agreement.

Once all these elements are incorporated into the contractual agreement, you can order the merchandise and schedule the installation. (In some states and instances, verbal agreements may be considered binding. But why risk such a tenuous position?) Written contracts offer you the greatest protection under the law and should be a part of every sale. Unforeseen things can go wrong in the course of any floor covering transaction, so it's always best to have a signed, written contract.

Purchasing

Astute purchasing can often provide you with a similar (or greater) percentage increase in profits than a similar percentage increase in sales. By buying a product at an amount that is 10 percent less than the usual wholesale price, when you sell that product at your regular retail price, you will make an additional 10 percent profit. To earn that same amount of additional gross profit through sales alone, you may have to increase your sales figures by as much as 20 percent. For a company to increase sales totals by 20 percent is no small feat. A tremendous amount of time and effort must be expended to achieve this goal. On the other hand, by keeping track of current special pricing promotions offered by the manufacturers, you can seize upon these opportunities to improve your bottom-line profit totals. This suggestion indicates how important the purchasing process is. It should never be taken for granted, and the hunt for the best possible deal should be an ongoing commitment.

The goal of the purchasing department is to obtain the best-quality product at the lowest cost with the most beneficial credit terms. When you can do this, you provide the best value to your customer and at the same time make the greatest profit for your company.

Once you have been in business for a while, you will see what types of products are good sellers. This will indicate the styles that should be included in your inventory. By having goods on hand that were purchased when the wholesale price was lowest, you can make the most profit and also have the opportunity to sell it to your customers with a promise of immediate installation. There will be no delay while you await delivery from the manufacturer. For a customer who is in a hurry, this is a big plus. Some customers who require such quick service might include landlords or property managers whose apartments have been vacated and require instantaneous refurbishing prior to the next occupancy. For such a customer, your ability to provide the right product at the right price and in the right time frame is what will establish loyalty to your company. The next time the customer needs to buy floor coverings, she or he will come right back to you.

Buying generic products also gives you a pricing edge. For example, an unbranded nylon carpet of equal face weight to a branded nylon carpet will have a lower wholesale price. By virtue of this fact, if you have clients who are not concerned about such distinctions, you can often increase your profits by stocking up on generic products because your overall markups can be greater.

Another way to increase your profit margin is to get the manufacturer to offer you *quantity discounts* when you purchase large amounts. If you have no problem storing merchandise, this plan could prove profitable. However, before you buy all kinds of stock, make sure that you can afford to tie up your capital in merchandise that may take some time to sell. A business that is just beginning may be better off purchasing a small amount of merchandise that can be moved quickly, while leaving the majority of the funds available for current expenses and operating costs.

The types of products you purchase will be determined by the kind of clients you have. Flooring companies that cater to property management trade, e.g., keep on hand rolls of inexpensive carpet so they can meet the requirements of lower prices and immediate installation. A higher-end store that has more discriminating clients may buy high-quality *closeouts* (merchandise discontinued by the manufacturers) at drastically reduced prices. Thus customers still get the luxurious quality they expect, but the seller makes a greater percentage of profit. Each party in the transaction is satisfied because their respective expectations were met.

The purchase of any product for sale that is not an inventory item is essentially a *custom order*. Since not every customer who comes into your store will like the products you have in stock, it will be necessary to order from the sample swatches you have available. When you calculate the wholesale cost for custom order products, be sure to add the special handling and freight charges. On large orders these costs can be substantial. If you fail to include these amounts in your price proposal to the client, the charges will have to be paid out of your pocket. Consequently, your profit for that particular job will be less than expected. If this happens too often, your overall profit margins will suffer and, to remain profitable, you will have to cut expenses in other areas to make up the difference.

When shipments arrive at your warehouse, they should be checked immediately for any visible signs of damage. In addition, all items should be counted and tallied to make sure the items delivered agree with those on the manifest. Any discrepancies should be immediately brought to the attention of the carrier who made the delivery. The person delivering the merchandise will require someone at the warehouse to sign a delivery receipt because this receipt is the carrier's proof that the merchandise was delivered in satisfactory condition. If the delivery receipt is signed and the materials prove to be defective at a later date, you could have a difficult time convincing the carrier of his or her responsibility. If any merchandise needs to be rejected due to any visible imperfections, make sure you have a written damage report signed by the driver to substantiate your claim. Furthermore, do not return any damaged goods to the manufacturer without written authorization.

If you receive, say, a roll of carpet that appears to be in pristine condition, only to discover once you roll it out that there is a visible flaw in it, you must notify the manufacturer at once. If it obviously has not been damaged in transit, then it is a manufacturer's defect. These types of defects could be anything from a black yarn streak on a white carpet, to a row of yarn that is tufted higher than the other rows. Unfortunately, these flaws are often not apparent until you are actually in the process of installing the carpet. This situation is among the worst nightmares of any flooring contractor. It can create a terrible inconvenience for all parties concerned. Tempers begin to heat up, and stress levels increase. It's at times like these that you need to keep your cool. Often, you will have to make certain concessions to your customers to keep them happy. In cases like this, you can only try to assure the customer of your intention to provide the best product possible and hope that the mill supports you in your time of need. However, no matter what, do not return any defective material to the manufacturer until you have the proper written authorization.

The manufacturer will issue you a *return authorization* (RA) number that must be attached to the merchandise when you ship it back. The mill wants this RA to accompany the returned merchandise for two reasons: first, to identify and inspect the goods; and second, to properly credit your account so you won't have to pay for the damaged product. If you were to unilaterally ship the unacceptable merchandise back to the manufacturer without proper authorization, since there's the possibility of the goods being lost or misplaced in transit, you would still be liable for payment because you initially received the delivery. No matter how inconvenient it may be to keep the damaged goods in your storage facility, do so and wait for the proper, written authorization forms from the manufacturer before you return anything.

Inventory control

Inventory control in the flooring business is just as critical as in any other retail operation. The need to keep track of your stock on hand is critical to your profitability. If merchandise is not adequately accounted for, losses through misuse or theft might go undetected. Whether your company stocks rolls of carpet, boxes of ceramic tile, or buckets of adhesive, a running record of what is on hand at any given moment should be consistently kept up to date. In order for you to fill orders to your customers on a timely basis, this information is vital.

Inventory control systems can be manual or carried out by computer. Software programs available today can track inventory in the blink of an eye. However, manual procedures may prove adequate for small companies that have limited stock on hand. Regardless of the method you chose, the best control system is a *perpetual inventory control system*. This procedure keeps an up-to-date record of all stock items. It notes the beginning inventory amount for a given product, less any deductions for sales and returns, thus providing you with the exact quantity amount on hand at that moment. It eliminates the need to constantly take physical inventories; yet actual counts should be taken periodically to verify that what is on hand equals the amount shown on the books. An annual or semiannual physical inventory may be sufficient for this purpose.

A satisfactory inventory control system will tell you not only what products are available but also what items are slow movers. In addition, it will alert you to the need to reorder certain products when they fall below a specified level. Items that are not selling well can have their quantities reduced, whereas rapid movers can have their amounts increased. Since inventory ties up so much of your working capital, it is important to stock merchandise that will turn over quickly. Sitting on, say, $1000 worth of merchandise that may take most of a year to totally sell is not a wise investment choice. Conversely, if you purchase a product that is in much greater demand and turn it over 6 times during that year, you've made a more worthwhile inventory decision.

Let's examine, for example, how you would inventory and keep track of a certain roll of carpet. The following scenario would take place regarding that roll while it is part of your inventory:

- It would be received in your warehouse and visually inspected to make sure there was no apparent damage.
- The roll tags from the mill would be verified as matching those on your purchase order for color, quantity, etc. If all details correspond, the delivery receipt may be signed and the item accepted.
- An inventory control card would be generated (Fig. 8-8). The information would be the same regardless of whether it's input on your computer or listed on a handwritten inventory sheet.
- After the original information was logged, each time a cut of any length were taken from the roll to fill an order, that deduction would be itemized on the inventory control card and the remaining balance would be shown.

A simple control card like Fig. 8-8 can be developed for each individual product you stock, but adjustments will have to be made with regard to the column headings. For instance, Cuts would not be a column heading for boxes of vinyl composition tile; it would just require Quantity Sold. It's often best to start out with blank column headings and develop titles from words that you really feel comfortable with. It's the idea of keeping track of your inventory that's important, not the column heading names.

Periodically check these cards to verify remaining balances, total cash amount invested in inventory, and reorder requirements. Avoiding an oversupply or undersupply of a given product is critical.

Employee management

The people who work for you are among your most valuable assets. Their enthusiasm and concern for the betterment of the company are imperative to its survival. Disgruntled employees or those who lack good customer service skills can ruin your store's reputation in no time. How you treat your employees and how they react to you will depend a great deal upon your leadership characteristics.

In a floor covering operation, there are basically two types of employees: office staff and installation specialists. The office staff will

Inventory Control Card

ICC#391

Manufacturer/ Customer	1 Date	2 Style	3 Color	4 Roll #	5 Size	6 Yardage Quantity	7 Square Footage Quantity	8 Cost Per Square Yard	9 Cost Per Square Foot	10 Cuts	11 Remaining Balances	12 Remaining Cost Value
1 XYZ Carpet Company	1/1/97	#1260 Magical	#427 Starburst	#107653	12 × 100$\underline{0}$	133⅓ sq. yd.	1200 sq. ft.	$5.$\underline{49}$	61¢	-0-	-133⅓ sq. yd. –1200 sq. ft	$731.$\underline{98}$
2 Mrs. Flynn– Inv#4361	1/20/97	"	"	"	"	⟨–55⅓⟩	⟨–498⟩	"	"	12 × 41$\underline{6}$	–78 sq. yd –702 sq. ft.	$303.$\underline{76}$
3 Mr. Sturm– Inv#4398	2/2/97	"	"	"	"	⟨–22⅓⟩	⟨–201⟩	"	"	12 × 16$\underline{9}$	–55⅔ sq. yd. –501 sq. ft.	$181.$\underline{17}$
4												
5												
6												
7												
8												

8-8 *Inventory control card.*

269

include secretaries, receptionists, bookkeepers, salespeople, ware-house workers, etc. Installation specialists (installers) are the individuals who are actually on the customers' premises installing the products purchased from your firm. Everyone on your staff has basic human needs and desires. Their values and attitudes reflect their fundamental psychological makeup. As you help them attain their own goals, they in turn help you fulfill yours.

Leadership characteristics

An effective leader interacts with others in such a way that they become motivated toward positive action. Through encouragement, guidance, and instruction, good leaders can influence others to do their jobs in a proficient manner. Some of the many qualities needed in a competent leader are patience, confidence, optimism, resourcefulness, open-mindedness, great communication skills, and the ability to solve problems. Of all these skills, the most important leadership quality for a flooring contractor is a problem-solving ability.

Often, when a problem arises, you will hear, "This is wrong, that is wrong, and I don't know how we'll ever fix it!" It is at this point that you must sit back, take a deep breath, and start evaluating the situation, intent upon finding a solution to the problem. Some people take so much pleasure in reporting bad news that they don't even think (or care) about investigating possible solutions to the problem. At these moments you must listen to what the problem entails in all its details and come up with the best, most cost-effective, customer-satisfying solution. The best solution is not always what is easiest for the installer or what costs the least; it's what is best for the customer. What will make the customer happy is what you have to strive for. Whatever will make the job look its finest is what is required. You have to cut through all the emotions, stress, and negativity of the moment, calm everything down, and say, "Okay, you've told me the problem. Now let's come up with a solution." Display your leadership qualities, and do what is right for both your company and your customer.

When you interact with your staff, always be honest and forthright. If they ever feel as if they have been lied to, they will sense that you can never be trusted. When that happens, it's difficult to be an effective leader. Human nature is such that if someone is caught in a lie, no matter how minor, people will think that all other statements by that person are lies, too. Furthermore, when someone lies, he or she has to continually reshape the tale by building it up with other lies. Before long, the story has radically changed, and the truth is nowhere to be found. Therefore, avoid such deceptive behavior at all costs by always telling the truth to your staff, no matter how painful it may be.

Office staff

Those individuals who work inside your office are truly instrumental in the success of your company. They are the ones with whom you spend most of your time. They are the backbone of your enterprise. They represent the image of your company to the public. If they are rude or unhelpful, you could lose many customers. If they are pleasant and courteous, sales will abound. Therefore, choosing the right employees is the first step toward success. The second step is to treat your employees well so they will continue to develop pleasing qualities.

Some characteristics to look for in employees are punctuality, loyalty, honesty, good personal hygiene, respect for authority, intelligence, and consistency. For the most part, these character attributes cannot be learned but are part of a person's basic personality. If one does not already possess them, it will be hard to acquire them as an adult. Consequently, in your preemployment interview process, you must be able to discern whether that person has these good qualities. You want to avoid hiring someone, spending time and money to train them, only to find that they lack the personality traits you sought. So, it's always a good idea to check any employment references on their written application. Identify yourself to the references, and see what kind of information they are willing to give you about the employment candidate.

Proper employee training is one of the best management tools available to you. Whenever possible, company policies should be in writing. This eliminates any misunderstandings and prevents employees from thinking they have to make up their own procedures. In addition, if ever there is a question whether a certain approach to a situation is appropriate, a quick reference to the policy statement will settle the matter. Each employee should be given a copy of the written policy handbook, and it should be updated whenever changes are made. Some of the items it should cover are as follows:

- Equal opportunity employer issues
- Performance reviews
- Promotion requirements
- Hours of employment
- Time card policies
- Lunch and break policies

- Work performance expectations
- Benefit plans—insurance, vacations, sick pay, etc.
- Company work rules
- Dress code
- Arbitration policies in case of disputes
- Training schedules
- Job descriptions
- Safety regulations

Remember, once you put these policies in writing, you must subscribe to them in total. Also, you must enforce them strictly and to all employees equally. You cannot play favorites or pick on someone unfairly. Legal proceedings could result if you do not treat all employees fairly and with equal respect. If you have to terminate an individual for poor performance or for being a disruptive influence in the working environment, document the corrective actions taken before dismissal and make that document a part of the employee's permanent file. Proper documentation can protect the company if any legal action is taken against you by a terminated employee.

In general, good training techniques, and an effort made by you to see to it that your employees are safe and secure in their jobs, will help your company run efficiently and profitably. Furthermore, your customers will sense this feeling of harmony and it will give them a good impression of your company.

Installation specialists

Installation specialists, also known as *mechanics* or *installers,* are a vital link in a floor covering company's chain of operations. This is the point in the entire sales transaction where you can have either complete success or utter failure. No matter at how low a price you have sold a particular product to a customer, if it is poorly installed, they will be disgusted and upset with the whole deal.

As we enter the 21st century, the biggest problem facing the flooring industry will be the availability of competent installation specialists. Despite the fact that there are some new schools of installation being developed in the country, there simply are not enough training facilities to properly teach students. Usually the only training that is available in the floor covering industry is a short 2-day or week-long seminar put on by flooring manufacturers. While these seminars are informative, they are really only geared to instruct the students on how to install that particular manufacturer's products. While that is understandable, it does not give the installers the well-rounded diversity they need to be productive today. Typically a good installer teaches assistants who, after some time, become proficient enough to perform installations on their own. Most flooring companies do not have the time or money necessary to train a floor covering installer. That process could take years before someone can be trusted to install flooring unsupervised. Furthermore, there is no guarantee that after they are sufficiently trained they will stay with that company. So, at best, it's a 50-50 proposition if you want to invest in someone's future. Therefore, you have to be very selective about the individuals you hire and what you can trust them to install as you learn their capabilities.

In the flooring industry it usually takes a minimum of 3 to 5 years before someone can be considered a qualified mechanic, before you, as a store owner or contractor, can entrust an expensive piece of merchandise into that person's care. Horror stories are told in the flooring community of installations gone bad due to an installer's incompetence. Care and attention to detail are the most important qualities necessary for a successful flooring installation. If installers merely want to "bang it in" so they can get paid for the job, the quality of workmanship will not be good. Unsightly seams and poorly laid out patterns are only a few of the things that can go wrong in a bad installation. Therefore, the best approach is to start slowly with a new installer. After you have interviewed the potential mechanic and seen that she or he has a supply of proper tools and a vehicle, start the installer out with small jobs that require minimal skills. Also after the installation, get the customer's opinion of the workmanship and the installer's demeanor and attitude. In addition, take the time to personally inspect the job. Look at it with a very critical eye to make sure everything is done to your satisfaction and to the level of professionalism you expect. As you become more familiar with the installer's work habits and performance, you can gradually assign increasingly more difficult jobs. Soon you will be able to give the installer your toughest jobs with confidence that the job will be done correctly and efficiently. After that, the installer will be a key member of your installation team.

Floor covering installers today are placed in a difficult position when they go to a job site. Not only are they required to be top-quality installation specialists, but also they must be furniture movers and customer relations personnel. It is no easy task, but since they are essentially the company's field representatives, they are an extension of your firm into the community. Therefore, they need to be advised and made well aware of their important role so they can behave accordingly. If they act unpleasant with a customer, it will reflect on your company. If they are helpful and polite, it will make the entire enterprise look good. Satisfied customers will tell their friends, and these kinds of

"I'm so happy" referrals are the very best form of advertisement. The greatest advertising campaign in the world cannot equal the impact of word-of-mouth referrals. A potential buyer who comes in on a referral is almost a guaranteed customer because someone else has done all the "leg work" for the buyer. The price has been shopped and the workmanship has been tested with quality results, so that by the time the referred customer comes in, she or he is virtually presold. All that is really necessary is to find out the customer's favorite flooring material, have the premises measured, and close the sale.

Due to the challenging conditions facing installers every day, the company owner must offer them the support they need. Not only should they be properly remunerated for their hard work, but also they should be confident that their efforts are appreciated. Far too often a job well done on a difficult installation goes unrecognized. Everyone has a need to have his or her personal worth validated. An occasional pat on the back can do wonders for an individual's self-esteem and enthusiasm for the work. So make an effort to let your employees know how highly valued they are and that their efforts and achievements are not unnoticed.

Insurance

As in any other endeavor in life, it's always best to protect yourself against unforeseen catastrophe. In the floor covering business, the best way to do this is to make sure you have adequate insurance coverage. The operative word here is *adequate*. Adequate coverage is not only a policy you can afford, but also a policy (or more than one policy) that protects you against occurrences which could most likely transpire in your type of business. The worst-case scenario, that once in a while does occur, is one in which you are the sole source of funds in the event of a major loss. If, say, there is a fire in your showroom/warehouse, the entire business burns down, and you have absolutely no fire insurance, that could be horrendous. The mere thought of that possibility is too awful to consider. When you begin to formulate your business plan and financial budget, identify and set aside sufficient funds to cover the many types of policies that you will need.

The major categories of insurance that most businesses need are property, casualty, vehicular, and life insurance. Always consult a professional insurance agent or broker for your insurance needs. An agent usually represents one company, whereas a broker handles insurance for many companies. Find an individual you can trust, one who will get you the best coverage and at the best rate for your budget.

A basic fire policy is property insurance to cover you against loss caused by fires and explosion. It can, however, be extended to include other perilous situations such as flood, earthquake, lightning strikes, installations, and accounts receivable losses. The premium amounts for a fire policy, as well as for the additional covered perils, are calculated by taking into consideration several factors including:

- Age of the building
- Type of construction materials that comprise the structure itself (i.e., the walls, ceiling, and floor makeup)
- Neighborhood where it is located
- Kinds of surrounding structures
- Amount of glass windows and doors in the building
- Current condition of the building

Payment to you by the insurance company for losses suffered can be calculated by one of two methods: actual cash value or replacement cost. *Actual cash value* is the value of the property at the time of the loss, which includes a reasonable deduction for depreciation. *Replacement cost* is the amount of money required to return the damaged or lost property to its former state. Your insurance agent should discuss these important features with you so that in the event you have a loss, you're not surprised by a settlement that is less than you thought it would be. Check to see if getting *replacement cost* loss payments coverage will increase your premium.

Liability insurance covers against bodily injuries caused to others as a result of your actions or negligence. When someone wishes to take legal action against you, this type of policy could help pay legal fees and damages if the incident is a peril that is included in your insurance coverage. A *general liability policy* will cover a great many occurrences, including acts of your employees while on the premises of your clients. (Certain restrictions may apply here, so examine your policy for the actual wording and coverage.)

Vehicle insurance, also a casualty policy, is a must for all vehicles within your organization. Bodily injury and property damage limits should be sufficiently high to cover expenses to any injured parties or damaged vehicles. Always go for the highest limits you can afford because hospital and doctor bills are very costly today. If someone is injured in an automobile accident, the cost of treatment could easily be many thousands of dollars. Furthermore, repairs or replacement of the vehicle would require substantial funding. Most states require proof of insurance coverage for both the driver and the vehicle, so be prepared for all possibilities by paying attention to the insurance coverage of your vehicles.

Life insurance on an owner or a partner helps to ensure a tranquil transition in the event of an untimely death. If the sole owner passes away, funds will be needed to continue the business. If the business is a partnership, the terms of the partnership agreement may have provided that a life insurance policy be maintained in the name of each partner in case of such eventuality. Whatever the reason, a businessperson's life insurance should be carefully reviewed during consultation with an insurance adviser.

Other types of insurance coverage could be included in your insurance package:

- Business interruption insurance
- Worker's compensation insurance
- Crime insurance

Business interruption insurance will cover such items as rent, employee salaries, or other major expenses if your building is severely damaged by a direct physical loss caused by a covered peril. Having this kind of coverage can be a true blessing if you ever have to use it; without it, you would have to pay all those expenses during the restoration period yourself. An example of these situations is often seen in the daily newspapers where we read about sudden flooding, a mudslide, or a plane falling from the sky. The unpredictable needs to be considered when you prepare your business for security.

Worker's compensation insurance is required by most states to protect employees against accidental injury and illnesses which result from their work. Be aware that negligence on the part of the employer is not a requirement for an employee to collect under a worker's compensation policy. Employees who file claims under this kind of policy can collect payments to cover lost wages and medical expenses. Premiums on worker's compensation policies are determined by the job classifications within your company and your total payroll amounts. For example, the premium amount due to cover a secretary's position will be much lower than that to cover a carpet layer's position because the physical activity and the propensity to injury are greater for a carpet layer than for a secretary. Therefore, the premium amount will vary from one job description to the next.

Crime coverage can include such incidents as theft, burglary, embezzlement, and pilferage. If these types of events are possible within your company, consider subscribing to this insurance coverage if the premiums are affordable. There are a lot of internal measures you can implement to try to avoid these situations, if the cost of the coverage is prohibitive. You should have most of these measures in place even if you can afford the insurance coverage. Yet, having the insurance can be another safety net in your insurance package.

A *business owner's package* (BOP) is an insurance policy designed for the small business owner that will provide a wide range of coverage against most common perils. It includes many of the coverages listed above such as general liability insurance, property insurance, non-owned vehicle insurance, and crime insurance. It is more cost-effective than purchasing separate policies for each category. It is designed to provide broad-based, low-cost coverage to any business owner in need of it.

Floor covering business and the future

As we approach the year 2000, anyone desiring to go into the floor covering business had better study hard and long to understand where the industry is today, and where it will be by the turn of the century. Currently, the entire flooring industry is in a complete state of flux. The changes taking place now are unprecedented. There is no telling what will happen next; yet predictions and speculations abound. Only one thing is certain, the landscape of the industry is changing at such an alarming rate that many companies operating today will have a difficult time surviving into the next millennium. If you truly want to enter the flooring business, it will take more than desire. It will take sufficient capital, a well-trained staff, and constant dedication. Many of the older concepts of operating a business—those with their roots in the 1950s and 1960s—are no longer applicable in the 1990s. It's a different marketplace with an entirely new set of rules.

One of the biggest changes taking place now affects the face of the competition. Your competition is no longer Mrs. Jenkins, two blocks over, who runs a little carpet shop with her husband. Now it's a huge home center that sells everything at a discount, from nails to appliances. The competition is also a *buying group* that independent store owners have subscribed to, in order to become part of a national purchasing organization. Having greater buying power, they can command better pricing from manufacturers. And finally, the competition is now the flooring manufacturers themselves, for they are entering the retail sector as well and wresting away your hard-fought-for market share. The pie chart is the same, but it's getting cut up in a completely different fashion and by fewer and fewer companies.

The entire business atmosphere in the United States is becoming more and more oligarchical. In an oligarchy, control is held in the hands of a few. We are seeing this, not only in the flooring industry, but also in most commercial sectors. As one company buys up another, and then yet another, that company's sphere of influence widens while the list of competitors shrinks. Once they have bought out most of their major competitors, all that is left is small, independently owned stores that pose little or no threat. This is taking place in most industries nationwide.

"Main Street, America" is a dying phenomenon. Many small stores, whether they be hardware stores, drug stores, ladies apparel shops, or sporting goods stores, find that often they cannot even buy the same product at the price the major chains are "selling" it for. It makes you wonder whether it's even worth trying to open your own business. But, the answer is yes; it is worth it. "But how?," you may ask.

Well, as stated earlier, it's worth it mainly from a self-esteem standpoint. No matter what type of business you choose, the satisfaction you receive from being your own boss is greater than can be expressed in words. Also, if you have any kind of entrepreneurial spirit, you will respond to whatever challenges you face and be successful because hard work and dedication will see you through. But most importantly, the person who can adapt to the changing business climates will be the one who will survive.

To adapt to this new marketplace, small dealers will need to rethink their mission statements. A *mission statement* is a company's philosophy of what it wants to achieve, i.e., its goal. A simple mission statement could be "to provide our clients with the most competitively priced products, along with the best possible customer service." These company goals should transfuse your entire business operation. From the receptionist, to the warehouse supervisor, to the company president, this mission statement should influence all transactions and decision making. By never losing sight of its purpose, a company grows into a more consistent unit. Customers respond to consistency because it gives them a feeling of security. Knowing that a company will stand behind its products and services allows the customer to purchase items from that company with confidence. And if the company does keep its promises, it can be assured of repeat customers and referrals.

For a company to persist, it will have to find a niche it can fill. Most consumerism today is price-driven. Loyalty to a specific company is a scarce commodity. When consumers can buy the same or similar product at a lower price elsewhere, in order to keep them at your store, you have to provide that "something special." You have to learn to compete on an entirely different level. It's unlikely you will be able to slug it out on a price basis only, so you will have to reinvent your company in order to be a viable force in the marketplace. Enormous capitalization is essential. Sustaining customer loyalty is something that must be worked at daily. Becoming a service-oriented business may be the only thing that separates you from the other stores, while still allowing you the luxury of a halfway decent markup on your products. Your reputation will be the key to your success. Treasure it as you would your most prized possession.

The uncharted waters that lay ahead are made more eerie by the way the flooring industry is dramatically reshaping itself. As a participant in that redefining plan, you will have to search for new directions that will reveal the strategies necessary for you to separate yourself from the crowd. By selling your clients what they want and by respecting their need for quality, convenience, and value, your company will be the one to flourish into the next century.

Resources

3M Company
  Specialty Chemical Division—
    3M Center
  Building 223 65 04
  St. Paul, MN 55144-1000

Aged Woods, Inc.
  2331 East Market Street
  York, PA 17402

American Plywood
  Association—The Engineered
    Wood Association
  PO Box 11700
  Tacoma, WA 98411-0700

Armstrong World Industries, Inc.
  PO Box 3001
  Lancaster, PA 17603

Bruce Hardwood Floors
  c/o The Bolton Group
  1806 Royal Lane
  Dallas, TX 75229

Carpet Cushion Council
  PO Box 546
  Riverside, CT 06878

Carpet and Rug Institute
  PO Box 2048
  Dalton, GA 30722

Crossley Carpet Company
  PO Box 745
  Truro B2N5G2
  Nova Scotia
  Canada

Dal Tile
  7834 Hawn Freeway
  Dallas, TX 75229

Du Pont Flooring Systems
    Company
  403 Holiday Avenue
  Dalton, GA 30720

Durango Trading Company
  602 The Eagle Pass
  Durango, CO 81301

IPOCORK, S.A.
  PO Box 13
  S Paio De Olieros
  4535 Lourosa
  Portugal

Kentucky Wood Floors
  4200 Reservoir Avenue
  Louisville, KY 40213

National Oak Flooring Manufac-
    turers Association
  PO Box 3009
  Memphis, TN 38173-0009

National Tile Contractors
Association
PO Box 13529
Jackson, MS 39236

Perstop Flooring Company
c/o Golin/Harris
500 North Michigan Avenue
Chicago, IL 50511

RCFI
c/o Stern & Associates
11 Commerce Drive
Cranford, NJ 07016

Roane Company
14141 Arbor Place
Cerritos, CA 90703-2464

Roberts Consolidated Industries
600 North Baldwin Park
Boulevard
Industry, CA 91749

Stanley Bostitch Company
PO Box 1739
East Grenwich, RI 02818

The Taunton Press
52 Church Hill Road, Box 355
Newtown, CT 06470

TEC Inc.
315 South Hicks Road
Palatine, IL 60067

Wellco Business Carpet
PO Box 281
Calhoun, GA 30701

Weyerhauser Company
2000 Frontis Plaza
Boulevard #101
Winston-Salem, NC 27103

Selected On-Line Resources for Floors and Floor Covering

see ONLINE RESOURCES FOR THE CONSTRUCTION INDUSTRY

Resources and links for people in the construction industry:
builders, contractors, construction managers, trades workers,
architects, designers, engineers, and vendors.
<http://207.76.230.67/constr.htm>

BuilderNet
BuilderNet index/Manufacturers/Design firms/Construction
companies/Associations and institutes/Library/BuilderNet Welcome
Center/ Add URL
<http:www.buildernet.com>

The World-Wide Web Virtual Library: Architecture—Construction
section
<http://www.clr.toronto.edu:1080/VIRTUALLIB/ARCH/mat.html>

Flooring subfloors and underlayment, ceramic tile, vinyl tile, sheet vinyl, installations, repairs, refinishing, and maintenance
<http://www.hometime.com>

Rejuvenate and repair stains on wooden floors, laying 3-in strip floor tiles
<http://www.housenet.com>

Replacement flooring, repairs, refinishing, ideas
<http://www.homeideas.com/planner>

Construction and installation of floors and subfloors
<http://www.apawood.org/buildertips>

Construction and installation of floors and subfloors
<http://www.umass.edu/bmatwt>

Kitchen floors, vinyl tile, staining floors
<http://www.teleport.com/howto>

Grouting, sealing, cleaning, maintenance
<http://www.aldonchem.com>

Articles on flooring
<http://www.homeownernet.com/articles>
<http://begin.com/fixit/struct.htm>

Glossary

adhesive—Liquefied substance used to secure various flooring products to a floor. Also known as *mastic*.

anchor bolt—A metal peg embedded into the top of a foundation wall, to which the sill plate is attached.

ANSI—American National Standards Institute.

APA–The Engineered Wood Association—Formerly known as the American Plywood Association, a trade organization that represents many U.S. wood panel manufacturers.

architectural drawings—Those drawings devised by an architect of a structure that show all the necessary information such as dimensions, elevations, floor plans, and details.

benchmark—A fixed identification marker on a piece of real property serving as a reference point on architectural drawings.

Berber—Currently most residential loop pile carpets are referred to as Berber carpets.

bid—See *proposal*.

bisque—The formation of raw materials into a ceramic tile form.

boilerplate—Certain portions of architectural drawings that are so common to most buildings that they are reproducible in their entirety from one structure to the next.

breadth—The width of any sheet flooring material, such as carpet or vinyl roll goods.

broadloom—Any carpet wider than 6 ft.

buttering—A process whereby flooring mastic is spread on not only the substrate, but also the back of a tile to ensure greater adhesion.

capillary action—A thin opening (vein), caused by surface tension, that allows moisture to pass through the soil and/or a concrete slab.

CBU—Cement backer board units.

CBD—Cash before delivery.

closeouts—First-quality merchandise that has been discontinued from the running line of a manufacturer.

COD—Cash on delivery.

combing—Spreading flooring mastic with a trowel into long ridges.

composition tile—VCT and luxury vinyl tile.

crazing—Hairline cracks on the surface of a ceramic tile or hard-surface flooring product.

CRI—Carpet and Rug Institute.

cross-seam—A seam that is perpendicular to a lengthwise edge of a sheet flooring material. It is created by attaching the width ends of the product together.

CSC—Construction Specifications Canada.

CSI—Construction Specification Institute.

cupping—A condition in which a wood board becomes either concave or convex across its face. This is caused by a moisture change on one side of the floor.

cushion—Any number of products used under carpets to provide stability and comfort; also known as *padding*.

delamination—Separation of the secondary carpet backing material (or attached cushion) from the primary backing material.

discount—The amount that can be deducted from payment of an invoice if the merchandise is paid for within a certain time; 2%/30 means a 2 percent discount if the amount is paid within 30 days of the invoice date.

do-it-yourselfer—An individual who wishes to personally install a product without the aid of a professional.

dry-laying—A technique used to preview how a tile floor will look before any adhesive is spread.

dynamic moisture condition—Water vapor that moves through, and is emitted from, a concrete slab.

edge ravel—Unraveling of the edges of a loop pile carpet.

embossing leveler—A cementitious mixture of portland cement and liquid latex that is spread over an existing resilient flooring to cover up an embossed texture.

environmentally friendly—Descriptor of products that do not contain harmful solvents or chemicals that are destructive to the environment.

EPA—Environmental Protection Agency.

fictitious business statement—When a business is operated under a name other than the legal name of the person or entity actually conducting the business, the name must be registered with local government authorities.

fill piece—Material used to fill in the remaining portion of a room after at least one breadth (width) of sheet floor material has been installed.

firing—The process of baking ceramic tiles in a kiln.

flash-coving—See *self-coving.*

friable—Quality of asbestos fibers that can be broken and dispersed into the atmosphere.

full-spread—A method used for installing felt-backed sheet vinyl flooring, whereby an adhesive is spread throughout the entire surface of the floor; unlike the perimeter installation method.

green slab—An uncured concrete slab.

greenware—An unfired ceramic tile bisque.

greige goods—Undyed carpet. In this state the carpet has been tufted to the primary backing, but it needs to be dyed and have the secondary backing applied.

gully—The distance between the wall and the tackless strip.

hard-surface flooring—Also known as resilient flooring, this group includes such products as linoleum, sheet vinyl flooring, vinyl composition tile, and luxury vinyl tile.

HEPA—high-efficiency particulate air.

hygroscopic—Able (such as wood) to readily take up and retain moisture.

inlaid sheet vinyl flooring—A process whereby chips of solid vinyl are compressed onto a backing material to create a sheet vinyl flooring product.

kilim—The flat weave of a rug.

mastic—See *adhesive*.

memory—The ability of a carpet fiber to regain its structure (bounce-back) after it has been compressed.

multimedia floors—A predominately wood floor that incorporates the use of other materials such as marble, metal, ceramic tile, or stone.

NOFMA—National Oak Flooring Manufacturers Association.

nonfriable—Condition in which a product that contains asbestos fibers remains bound to its original backing material and has not been crumbled, crushed, or sanded.

OSB—Oriented-strand board.

overwood—A circumstance where a piece of wood may be slightly higher than an adjacent piece, thereby creating an uneven surface.

padding—See *cushion*.

peaking—Two edges of a seam begin to rise and form what appears to be a mountainous peak.

perimeter installation method—A technique used for installing vinyl-backed sheet vinyl flooring in which a 4- to 6-in band of adhesive is spread around the perimeter of the room and at the seams; unlike the full-spread method.

principals—Individuals who are the leading or controlling authorities in an activity. In a partnership, the partners are the principals.

proposal—Compilation of all necessary data and prices on a given project for consideration by the owner; also referred to as a *bid* or *takeoff*.

PVC—Polyvinyl cloride.

quarter turn—The act of turning a flooring product 45° to an already laid width or piece of the same material.

RA—Return authorization.

recessed scribe—A process of scoring a line, with a sharp metal needle, onto an overlapping edge of a sheet good product to delineate where the seam should be cut.

RFCI—Resilient Floor Covering Institute.

ripping—The act of cutting a piece of strip wood lengthwise, with the grain, to fit in a certain section of an installation.

rotogravure—A process whereby sheet vinyl flooring is decorated by printing a pattern on a base of vinyl and foam.

scale—A unit of measurement used for architectural drawings whereby a fractional amount is used to represent a greater amount, say, $\frac{1}{4}$ in equals 1 ft. Also, an actual ruler used to interpret the scaled distances on a set of architectural drawings.

seam contamination—A circumstance whereby excess adhesive that is used to adhere a resilient flooring product to a substrate, enters the seam.

self-coving—An installation method whereby sheet vinyl flooring is extended from the floor up the wall to a height of about 4 to 6 in; also known as *flash-coving*.

set of plans—The complete written documents that pertain to a given structure. This set includes both the architectural drawings and the written specifications.

shop drawings—Those drawings provided by the floor covering contractor to the installer, or any necessary party, that accurately depict all the necessary information, and room drawings, regarding a given job.

shot—The length cut of a piece of sheet flooring material, such as carpet or vinyl roll goods.

sill plate—A framing member that is attached to the top of the foundation on which rests the floor system and wall frame system.

skin-over—A film that develops on the top of a ceramic tile adhesive after it has been exposed for too long a period of time.

solution-dyed yarns—A process whereby color is added to yarn in its liquid state, thus becoming an integral part of the fiber.

specifications—Written instructions that supplement the information contained on the architectural drawings.

static moisture condition—The actual water present in a concrete slab.

stay-tacking—The temporary securing of carpet by nailing certain sections of an installation.

straight-edge—A long metal tool that has truly straight edges (usually 6 or 12 ft long) used to cut seams on carpet or other flooring materials.

subfloor—The underlying surface of a floor; also referred to as *substrate*.

substrate—See *subfloor*.

tackless strip—The 1-in wood strips that contain several rows of sharp, angled metal pins onto which carpet is stretched and secured.

takeoff—See *proposal*.

TCA—Tile Council of America.

telegraphing—The visual emergence of a product or substance through another.

thin-set—Cement-based powders that are combined with a liquid to create a mortar-based adhesive for setting ceramic tiles.

time and materials—Also known as "T & M," a term that refers to the amount of time and the cost of all materials necessary to complete an entire job or a certain phase of a job. The T & M proposal is in contrast to a set-price bid where the total price of a job is known from the onset of the project.

Trade credit—Credit terms offered by suppliers to their clients whereby payment of the invoice for goods is delayed a certain length of time, typically 30 days, i.e., net/30.

transposition—Change or alteration in the relative order or sequence of numbers in a series, say, writing 123 as 132.

tuft bind—The amount of force required to pull a tuft of yarn from the backing of a carpet.

UCI—Uniform construction index.

VAT—Vinyl asbestos tile.

VCT—Vinyl composition tile.

VOC—Volatile organic compounds.

warp—Lengthwise strand of yarn used as a base for woven carpets.

weft—Widthwise yarn interlaced with warp yarn to weave a rug or carpet.

Index

About the Author

Peter Fleming has been in the floorcovering industry since 1976. He has operated his own full-service company in Santa Barbara, California, for the past 18 years. He also owns a recently opened oriental rug store that handles a wide variety of fine quality area rugs.

Mr. Fleming oversees every aspect of the flooring business, from sales to purchasing to accounting. Known for his ability to properly analyze on-site conditions and take accurate measurements, his expertise has been forged by years of hard work and dedication.

Mr. Fleming has a Bachelor of Arts degree in English Literature from Southern Illinois University. He is married and has one child. Active in community organizations, he is past-president of the Northside Optimist Club as well as past-president of the Greater Eastside Merchants Association of Santa Barbara.

Mr. Fleming is a licensed California State Contractor. His license is in classification C-15–Flooring. He is also a licensed Real Estate Agent.